Familiar Letters on Chemistry

By
Freiherr Justus von Liebig

20
EasyRead Super Large

RHYW
ReadHowYouWant Classics Library

ReadHowYouWant
YOUR CUSTOMIZED BOOK SOURCE

ReadHowYouWant provides access to great literature for all readers. We are committed to bringing you the largest library of classic literature and to making reading as effortless as possible.

We offer high-quality editions of the complete works of all the major literary authors, plus critical books about each author's life and works. Various editions of many titles are available, and for texts not written in English, we are often able to offer a choice of translations.

Our goal is to make reading easier, more enjoyable, and accessible to all readers. And for readers who need alternate format books, we provide the same selection in DAISY, MP3, eBook, and Braille.

ReadHowYouWant also allows readers to customize the books they choose — see ReadHowYouWant.com/customize

For more information please visit ReadHowYouWant.com/format

TABLE OF CONTENTS

FAMILIAR LETTERS ON CHEMISTRY.

AND ITS RLEATION TO COMMERCE, PHYSIOLOGY, AND AGRICULTURE,

Preface

The Letters contained in this little Volume embrace some of the most important points of the science of Chemistry, in their application to Natural Philosophy, Physiology, Agriculture, and Commerce. Some of them treat of subjects which have already been, or will hereafter be, more fully discussed in my larger works. They were intended to be mere sketches, and were written for the especial purpose of exciting the attention of governments, and an enlightened public, to the necessity of establishing Schools of Chemistry, and of promoting, by every means, the study of a science so intimately connected with the arts, pursuits, and social wellbeing of modern civilised nations.

For my own part I do not scruple to avow the conviction, that ere long, a knowledge of the principal truths of Chemistry will be expected in every educated man, and that it will be as necessary to the Statesman, the Political

Economist, and the Practical Agriculturist, as it is already indispensable to the Physician, and the Manufacturer.

In Germany, such of these Letters as have been already published, have not failed to produce some of the results anticipated. New professorships have been established in the Universities of Goettingen and Wuertzburg, for the express purpose of facilitating the application of chemical truths to the practical arts of life, and of following up the new line of investigation and research – the bearing of Chemistry upon Physiology, Medicine, and Agriculture, – which may be said to be only just begun.

My friend, Dr. Ernest Dieffenbach, one of my first pupils, who is well acquainted with all the branches of Chemistry, Physics, Natural History, and Medicine, suggested to me that a collection of these Letters would be acceptable to the English public, which has so favourably received my former works.

I readily acquiesced in the publication of an English edition, and undertook to write a few additional Letters, which should embrace some conclusions I have arrived at, in my recent investigations, in connection with the application of chemical science to the physiology of plants and agriculture.

My esteemed friend, Dr. Gardner, has had the kindness to revise the manuscript and the proof sheets for publication, for which I cannot refrain expressing my best thanks.

It only remains for me to add a hope, that this little offering may serve to make new friends to our beautiful and useful science, and be a remembrancer to those old friends who have, for many years past, taken a lively interest in all my labours.

LETTER I

My dear Sir,

The influence which the science of chemistry exercises upon human industry, agriculture, and commerce; upon physiology, medicine, and other sciences, is now so interesting a topic of conversation everywhere, that it may be no unacceptable present to you if I trace in a few familiar letters some of the relations it bears to these various sciences, and exhibit for you its actual effect upon the present social condition of mankind.

In speaking of the present state of chemistry, its rise and progress, I shall need no apology if, as a preliminary step, I call your attention to the implements which the chemist employs – the means which are indispensable to his labours and to his success.

These consist, generally, of materials furnished to us by nature, endowed with many most remarkable properties fitting them for our purposes; if one of them is a production of art, yet its adaptation to the use of mankind, – the qualities which render it available to us, – must be referred to the same source as those derived immediately from nature.

Cork, Platinum, Glass, and Caoutchouc, are the substances to which I allude, and which minister so essentially to modern chemical investigations. Without them, indeed, we might have made some progress, but it would have been slow; we might have accomplished much, but it would have been far less than has been done with their aid. Some persons, by the employment of expensive substances, might have successfully pursued the science; but incalculably fewer minds would have been engaged in its advancement. These materials have only been duly appreciated and fully adopted within a very recent period. In the time of Lavoisier, the rich alone could make chemical researches; the necessary apparatus could only be procured at a very great expense.

And first, of Glass: every one is familiar with most of the properties of this curious substance; its transparency, hardness, destitution of colour, and stability under ordinary circumstances: to these obvious qualities we may add those which especially adapt it to the use of the chemist, namely, that it is unaffected by most acids or other fluids contained within it. At certain temperatures it becomes more ductile and plastic than wax, and may be made to assume in our hands, before the flame of a common lamp, the form of every vessel we need to contain our materials, and of every apparatus required to pursue our experiments.

Then, how admirable and valuable are the properties of Cork! How little do men reflect upon the inestimable worth of so common a substance! How few rightly esteem the importance of it to the progress of science, and the moral advancement of mankind! – There is no production of nature or art equally adapted to the purposes to which the chemist applies it. Cork consists of a soft, highly elastic substance, as a basis, having diffused throughout a matter with properties resembling wax, tallow, and resin, yet dissimilar to all of these, and termed suberin. This renders it perfectly impermeable to fluids, and, in a great measure, even to gases. It is thus the fittest material we possess for closing our bottles, and retaining their contents. By its means, and with the aid of Caoutchouc, we connect our vessels and tubes of glass, and construct the most complicated apparatus. We form joints and links of connexion, adapt large apertures to small, and thus dispense altogether with the aid of the brassfounder and the mechanist. Thus the implements of the chemist are cheaply and easily procured, immediately adapted to any purpose, and readily repaired or altered.

Again, in investigating the composition of solid bodies, – of minerals, – we are under the necessity of bringing them into a liquid state, either by solution or fusion. Now vessels of glass, of porcelain, and of all non-metallic substances, are destroyed by the means we employ for that purpose, – are acted upon by many acids, by alkalies and the alkaline carbonates. Crucibles of gold and silver would melt at high temperatures. But we have a combination of all the qualities we can desire in Platinum. This metal was only first adapted to these uses about fifty years since. It is cheaper than gold, harder and more durable than silver, infusible at all temperatures of our furnaces, and is left intact by acids and alkaline carbonates. Platinum unites all the valuable properties of gold and of porcelain, resisting the action of heat, and of almost all chemical agents.

As no mineral analysis could be made perfectly without platinum vessels, had we not possessed this metal, the composition of minerals would have yet remained unknown; without cork and caoutchouc we should have required the costly aid of the mechanician at every step. Even without the latter of these adjuncts our instruments would have been far more costly and fragile. Possessing all these gifts of nature, we economise incalculably our time – to us more precious than money!

Such are our instruments. An equal improvement has been accomplished in our laboratory. This is no longer the damp, cold, fireproof vault of the metallurgist, nor the manufactory of the druggist, fitted up with stills and retorts. On the contrary, a light, warm, comfortable room, where beautifully constructed lamps supply the place of furnaces, and the pure and odourless flame of gas, or of spirits of wine, supersedes coal and other fuel, and gives us all the fire we need; where health is not invaded, nor the free exercise of thought impeded: there we pursue our inquiries, and interrogate Nature to reveal her secrets.

To these simple means must be added "The Balance," and then we possess everything which is required for the most extensive researches.

The great distinction between the manner of proceeding in chemistry and natural philosophy is, that one weighs, the other measures. The natural philosopher has applied his measures to nature for many centuries, but only for fifty years have we attempted to advance our philosophy by weighing.

For all great discoveries chemists are indebted to the "balance" – that incomparable instrument which gives permanence to every observation, dispels all ambiguity, establishes truth, detects error, and guides us in the true path of inductive science.

The balance, once adopted as a means of investigating nature, put an end to the school of Aristotle in physics. The explanation of natural phenomena by mere fanciful speculations, gave place to a true natural philosophy. Fire, air, earth, and water, could no longer be regarded as elements. Three of them could henceforth be considered only as significative of the forms in which all matter exists. Everything with which we are conversant upon the surface of the earth is solid, liquid, or aeriform; but the notion of the elementary nature of air, earth, and water, so universally held, was now discovered to belong to the errors of the past.

Fire was found to be but the visible and otherwise perceptible indication of changes proceeding within the, so called, elements.

Lavoisier investigated the composition of the atmosphere and of water, and studied the many wonderful offices performed by an element common to both in the scheme of nature, namely, oxygen: and he discovered many of the properties of this elementary gas.

After his time, the principal problem of chemical philosophers was to determine the composition of the solid matters composing the earth. To the eighteen metals previously known were soon added twenty-four discovered to be constituents of minerals. The great mass of the earth was shown to be composed of metals in combination with oxygen, to which they are united in one, two, or more definite and unalterable proportions, forming compounds which are termed metallic oxides, and these, again, combined with oxides of other bodies, essentially different to metals, namely, carbon and silicium. If to these we add certain compounds of sulphur with metals, in which the sulphur takes the place of oxygen, and forms sulphurets, and one other body, – common salt, – (which is a compound of sodium and chlorine), we have every substance which exists in a solid form upon our globe in any very considerable mass. Other compounds, innumerably various, are found only in small scattered quantities.

The chemist, however, did not remain satisfied with the separation of minerals into their component elements, i.e. their analysis; but he sought by synthesis, i.e. by combining the separate elements and forming substances similar to those constructed by nature, to prove the accuracy of his processes and the correctness of his conclusions. Thus he formed, for instance, pumice-stone, feldspar, mica, iron pyrites, artificially.

But of all the achievements of inorganic chemistry, the artificial formation of lapis lazuli was the most brilliant and the most conclusive. This mineral, as presented to us by nature, is calculated powerfully to arrest our attention by its beautiful azure-blue colour, its remaining unchanged by exposure to air or to fire, and furnishing us with a most valuable pigment, Ultramarine, more precious than gold!

The analysis of lapis lazuli represented it to be composed of silica, alumina, and soda, three colourless bodies, with sulphur and a trace of iron. Nothing could be discovered in it of the nature of a pigment, nothing to which its blue colour could be referred, the cause of which was searched for in vain. It might therefore have been supposed that the analyst was here altogether at fault, and that at any rate its artificial production must be impossible. Nevertheless, this has been accomplished, and simply by combining in the proper proportions, as determined by analysis, silica, alumina, soda, iron, and sulphur. Thousands of pounds weight are now manufactured from these ingredients, and this artificial ultramarine is as beautiful as the natural, while for the price of a single ounce of the latter we may obtain many pounds of the former.

With the production of artificial lapis lazuli, the formation of mineral bodies by synthesis ceased to be a scientific problem to the chemist; he has no longer sufficient interest in it to pursue the subject. He may now be satisfied that analysis will reveal to him the true constitution of minerals. But to the mineralogist and geologist it is still in a great measure an unexplored field, offering inquiries of the highest interest and importance to their pursuits.

After becoming acquainted with the constituent elements of all the substances within our reach and the mutual relations of these elements, the remarkable transmutations to which the bodies are subject under the influence of the vital powers of plants and animals, became the principal object of chemical investigations, and the highest point of interest. A new science, inexhaustible as life itself, is here presented us, standing upon the sound and solid foundation of a well established inorganic chemistry. Thus the progress of science is, like the development of nature's works, gradual and expansive. After the buds and branches spring forth the leaves and blossoms, after the blossoms the fruit.

Chemistry, in its application to animals and vegetables. endeavours jointly with physiology to enlighten us respecting the mysterious processes and sources of organic life.

LETTER II

My dear Sir,

In my former letter I reminded you that three of the supposed elements of the ancients represent the forms or state in which all the ponderable matter of our globe exists; I would now observe, that no substance possesses absolutely any one of those conditions; that modern chemistry recognises nothing unchangeably solid, liquid, or aeriform: means have been devised for effecting a change of state in almost every known substance. Platinum, alumina, and rock crystal, it is true, cannot be liquified by the most intense heat of our furnaces, but they melt like wax before the flame of the oxy-hydrogen blowpipe. On the other hand, of the twenty-eight gaseous bodies with which we are acquainted, twenty-five may be reduced to a liquid state, and one into a solid. Probably, ere long, similar changes of condition will be extended to every form of matter.

There are many things relating to this condensation of the gases worthy of your attention. Most aeriform bodies, when subjected to compression, are made to occupy a space which diminishes in the exact ratio of the increase of the compressing force. Very generally, under a force double or triple of the ordinary atmospheric pressure, they become one half or one third their former volume. This was a long time considered to be a law, and known as the law of Marriotte; but a more accurate study of the subject has demonstrated that this law is by no means of general application. The volume of certain gases does not decrease in the ratio of the increase of the force used to compress them, but in some, a diminution of their bulk takes place in a far greater degree as the pressure increases.

Again, if ammoniacal gas is reduced by a compressing force to one-sixth of its volume, or carbonic acid is reduced to one thirty-sixth, a portion of them loses entirely the form of a gas, and becomes a liquid, which, when the pressure is withdrawn, assumes again in an instant its gaseous state – another deviation from the law of Marriotte.

Our process for reducing gases into fluids is of admirable simplicity. A simple bent tube, or a reduction of temperature by artificial means, have superseded the powerful compressing machines of the early experimenters.

The cyanuret of mercury, when heated in an open glass tube, is resolved into cyanogen gas and metallic mercury; if this substance is heated in a tube hermetically sealed, the decomposition occurs as before, but the gas, unable to escape, and shut up in a space several hundred times smaller than it would occupy as gas under the ordinary atmospheric pressure, becomes a fluid in that part of the tube which is kept cool.

When sulphuric acid is poured upon limestone in an open vessel, carbonic acid escapes with effervescence as a gas, but if the decomposition is effected in a strong, close, and suitable vessel of iron, we obtain the carbonic acid in the state of liquid. In this manner it may be obtained in considerable quantities, even many pounds weight. Carbonic acid is separated from other bodies with which it is combined as a fluid under a pressure of thirty-six atmospheres.

The curious properties of fluid carbonic acid are now generally known. When a small quantity is permitted to escape into the atmosphere, it assumes its gaseous state with extraordinary rapidity, and deprives the remaining fluid of caloric so rapidly that it congeals into a white crystalline mass like snow: at first, indeed, it was thought to be really snow, but upon examination it proved to be pure frozen carbonic acid. This solid, contrary to expectation, exercises only a feeble pressure upon the surrounding medium. The fluid acid inclosed in a glass tube rushes at once, when opened, into a gaseous state, with an explosion which shatters the tube into fragments; but solid carbonic acid can be handled without producing any other effect than a feeling of intense cold. The particles of the carbonic acid being so closely approximated in the solid, the whole force of cohesive attraction (which in the fluid is weak) becomes exerted, and opposes its tendency to assume its gaseous state; but as it receives heat from surrounding bodies, it passes into gas gradually and without violence. The transition of solid carbonic acid into gas deprives all around it of caloric so rapidly and to so great an extent, that a degree of cold is produced immeasurably great, the greatest indeed known. Ten, twenty, or more pounds weight of mercury, brought into contact with a mixture of ether and solid carbonic acid, becomes in a few moments firm and malleable. This, however, cannot be accomplished without considerable danger. A melancholy accident occurred at Paris, which will probably prevent for the future the formation of solid carbonic acid in these large quantities, and deprive the next generation of the gratification of witnessing these curious experiments. Just before the commencement of the lecture in the Laboratory of the Polytechnic School, an iron cylinder, two feet and a half long and one foot in diameter, in which carbonic acid had been developed for experiment before the class, burst, and its fragments were scattered about with the most tremendous force; it cut off both the legs of the assistant and killed him on the spot. This vessel, formed of the strongest cast-iron, and shaped like a cannon, had often been employed to exhibit experiments in the presence of the students. We can scarcely think, without shuddering, of the dreadful calamity such an explosion would have occasioned in a hall filled with spectators.

When we had ascertained the fact of gases becoming fluid under the influence of cold or pressure, a curious property possessed by charcoal, that of absorbing gas to the extent of many times its volume, – ten, twenty, or even as in the case of ammoniacal gas or muriatic acid gas, eighty or ninety fold, – which had been long known, no longer remained a mystery. Some gases are absorbed and condensed within the pores of the charcoal, into a space several hundred times smaller than they before occupied; and there is now no doubt they there become fluid, or assume a solid state. As in a thousand other instances, chemical action here supplants mechanical forces. Adhesion or heterogeneous attraction, as it is termed, acquired by this discovery a more extended meaning; it had never before been thought of as a cause of change of state in matter; but it is now evident that a gas adheres to the surface of a solid body by the same force which condenses it into a liquid.

The smallest amount of a gas, – atmospheric air for instance, – can be compressed into a space a thousand times smaller by mere mechanical pressure, and then its bulk must be to the least measurable surface of a solid body, as a grain of sand to a mountain. By the mere effect of mass, – the force of gravity, – gaseous molecules are attracted by solids and adhere to their surfaces; and when to this physical force is added the feeblest chemical affinity, the liquifiable gases cannot retain their gaseous state. The amount of air condensed by these forces upon a square inch of surface is certainly not measurable; but when a solid body, presenting several hundred square feet of surface within the space of a cubic inch, is brought into a limited volume of gas, we may understand why that volume is diminished, why all gases without exception are absorbed. A cubic inch of charcoal must have, at the lowest computation, a surface of one hundred square feet. This property of absorbing gases varies with different kinds of charcoal: it is possessed in a higher degree by those containing the most pores, i.e. where the pores are finer; and in a lower degree in the more spongy kinds, i.e. where the pores are larger.

In this manner every porous body – rocks, stones, the clods of the fields, air, and therefore oxygen; the smallest solid molecule is thus surrounded by its own atmosphere of condensed oxygen; and if in their vicinity other bodies exist which have an affinity for oxygen, a combination is effected. When, for instance, carbon and hydrogen are thus present, they are converted into nourishment for vegetables, – into carbonic acid and water. The development of heat when air is imbibed, and the production of steam when the earth is moistened by rain, are acknowledged to be consequences of this condensation by the action of surfaces.

But the most remarkable and interesting case of this kind of action is the imbibition of oxygen by metallic platinum. This metal, when massive, is of a lustrous white colour, but it may be brought, by separating it from its solutions, into so finely divided a state, that its particles no longer reflect light, and it forms a powder as black as soot. In this condition it absorbs eight hundred times its volume of oxygen gas, and this oxygen must be contained within it in a state of condensation very like that of fluid water.

When gases are thus condensed, i.e. their particles made to approximate in this extraordinary manner, their properties can be palpably shown. Their chemical actions become apparent as their physical characteristic disappears. The latter consists in the continual tendency of their particles to separate from each other; and it is easy to imagine that this elasticity of gaseous bodies is the principal impediment to the operation of their chemical force; for this becomes more energetic as their particles approximate. In that state in which they exist within the pores or upon the surface of solid bodies, their repulsion ceases, and their whole chemical action is exerted. Thus combinations which oxygen cannot enter into, decompositions which it cannot effect while in the state of gas, take place with the greatest facility in the pores of platinum containing condensed oxygen. When a jet of hydrogen gas, for instance, is thrown upon spongy platinum, it combines with the oxygen condensed in the interior of the mass; at their point of contact water is formed, and as the immediate consequence heat is evolved; the platinum becomes red hot and the gas is inflamed. If we interrupt the current of the gas, the pores of the platinum become instantaneously filled again with oxygen; and the same phenomenon can be repeated a second time, and so on interminably.

In finely pulverised platinum, and even in spongy platinum, we therefore possess a perpetuum mobile – a mechanism like a watch which runs out and winds itself up – a force which is never exhausted – competent to produce effects of the most powerful kind, and self-renewed ad infinitum.

Many phenomena, formerly inexplicable, are satisfactorily explained by these recently discovered properties of porous bodies. The metamorphosis of alcohol into acetic acid, by the process known as the quick vinegar manufacture, depends upon principles, at a knowledge of which we have arrived by a careful study of these properties.

LETTER III

My dear Sir,

The manufacture of soda from common culinary salt, may be regarded as the foundation of all our modern improvements in the domestic arts; and we may take it as affording an excellent illustration of the dependence of the various branches of human industry and commerce upon each other, and their relation to chemistry.

Soda has been used from time immemorial in the manufacture of soap and glass, two chemical productions which employ and keep in circulation an immense amount of capital. The quantity of soap consumed by a nation would be no inaccurate measure whereby to estimate its wealth and civilisation. Of two countries, with an equal amount of population, the wealthiest and most highly civilised will consume the greatest weight of soap.

This consumption does not subserve sensual gratification, nor depend upon fashion, but upon the feeling of the beauty, comfort, and welfare, attendant upon cleanliness; and a regard to this feeling is coincident with wealth and civilisation. The rich in the middle ages concealed a want of cleanliness in their clothes and persons under a profusion of costly scents and essences, whilst they were more luxurious in eating and drinking, in apparel and horses. With us a want of cleanliness is equivalent to insupportable misery and misfortune.

Soap belongs to those manufactured products, the money value of which continually disappears from circulation, and requires to be continually renewed. It is one of the few substances which are entirely consumed by use, leaving no product of any worth. Broken glass and bottles are by no means absolutely worthless; for rags we may purchase new cloth, but soap-water has no value whatever. It would be interesting to know accurately the amount of capital involved in the manufacture of soap; it is certainly as large as that employed in the coffee trade, with this important difference as respects Germany, that it is entirely derived from our own soil.

France formerly imported soda from Spain, – Spanish sodas being of the best quality – at an annual expenditure of twenty to thirty millions of francs. During the war with England the price of soda, and consequently of soap and glass, rose continually; and all manufactures suffered in consequence.

The present method of making soda from common salt was discovered by Le Blanc at the end of the last century. It was a rich boon for France, and became of the highest importance during the wars of Napoleon. In a very short time it was manufactured to an extraordinary extent, especially at the seat of the soap manufactories. Marseilles possessed for a time a monopoly of soda and soap. The policy of Napoleon deprived that city of the advantages derived from this great source of commerce, and thus excited the hostility of the population to his dynasty, which became favourable to the restoration of the Bourbons. A curious result of an improvement in a chemical manufacture! It was not long, however, in reaching England.

In order to prepare the soda of commerce (which is the carbonate) from common salt, it is first converted into Glauber's salt (sulphate of soda). For this purpose 80 pounds weight of concentrated sulphuric acid (oil of vitriol) are required to 100 pounds of common salt. The duty upon salt checked, for a short time, the full advantage of this discovery; but when the Government repealed the duty, and its price was reduced to its minimum, the cost of soda depended upon that of sulphuric acid.

The demand for sulphuric acid now increased to an immense extent; and, to supply it, capital was embarked abundantly, as it afforded an excellent remuneration. the origin and formation of sulphuric acid was studied most carefully; and from year to year, better, simpler, and cheaper methods of making it were discovered. With every improvement in the mode of manufacture, its price fell; and its sale increased in an equal ratio.

Sulphuric acid is now manufactured in leaden chambers, of such magnitude that they would contain the whole of an ordinary-sized house. As regards the process and the apparatus, this manufacture has reached its acme – scarcely is either susceptible of improvement. The leaden plates of which the chambers are constructed, requiring to be joined together with lead (since tin or solder would be acted on by the acid), this process was, until lately, as expensive as the plates themselves; but now, by means of the oxy-hydrogen blowpipe, the plates are cemented together at their edges by mere fusion, without the intervention of any kind of solder.

And then, as to the process: according to theory, 100 pounds weight of sulphur ought to produce 306 pounds of sulphuric acid; in practice 300 pounds are actually obtained; the amount of loss is therefore too insignificant for consideration.

Again; saltpetre being indispensable in making sulphuric acid, the commercial value of that salt had formerly an important influence upon its price. It is true that 100 pounds of saltpetre only are required to 1000 pounds of sulphur; but its cost was four times greater than an equal weight of the latter.

Travellers had observed near the small seaport of Yquiqui, in the district of Atacama, in Peru, an efflorescence covering the ground over extensive districts. This was found to consist principally of nitrate of soda. Advantage was quickly taken of this discovery. The quantity of this valuable salt proved to be inexhaustible, as it exists in beds extending over more than 200 square miles. It was brought to England at less than half the freight of the East India saltpetre (nitrate of potassa); and as, in the chemical manufacture neither the potash nor the soda were required, but only the nitric acid, in combination with the alkali, the soda-saltpetre of South America soon supplanted the potash-nitre of the East. The manufacture of sulphuric acid received a new impulse; its price was much diminished without injury to the manufacturer; and, with the exception of fluctuations caused by the impediments thrown in the way of the export of sulphur from Sicily, it soon became reduced to a minimum, and remained stationary.

Potash-saltpetre is now only employed in the manufacture of gunpowder; it is no longer in demand for other purposes; and thus, if Government effect a saving of many hundred thousand pounds annually in gunpowder, this economy must be attributed to the increased manufacture of sulphuric acid.

We may form an idea of the amount of sulphuric acid consumed, when we find that 50,000 pounds weight are made by a small manufactory, and from 200,000 to 600,000 pounds by a large one annually. This manufacture causes immense sums to flow annually into Sicily. It has introduced industry and wealth into the arid and desolate districts of Atacama. It has enabled us to obtain platina from its ores at a moderate and yet remunerating price; since the vats employed for concentrating this acid are constructed of this metal, and cost from 1000l. to 2000l. sterling. It leads to frequent improvements in the manufacture of glass, which continually becomes cheaper and more beautiful. It enables us to return to our fields all their potash – a most valuable and important manure – in the form of ashes, by substituting soda in the manufacture of glass and soap.

It is impossible to trace, within the compass of a letter, all the ramifications of this tissue of changes and improvements resulting from one chemical manufacture; but I must still claim your attention to a few more of its most important and immediate results. I have already told you, that in the manufacture of soda from culinary salt, it is first converted into sulphate of soda. In this first part of the process, the action of sulphuric acid produces muriatic acid to the extent of one-and-a-half the amount of the sulphuric acid employed. At first, the profit upon the soda was so great, that no one took the trouble to collect the muriatic acid: indeed it had no commercial value. A profitable application of it was, however, soon discovered: it is a compound of chlorine, and this substance may be obtained from it purer than from any other source. The bleaching power of chlorine has long been known; but it was only employed upon a large scale after it was obtained from this residuary muriatic acid, and it was found that in combination with lime it could be transported to distances without inconvenience. Thenceforth it was used for bleaching cotton; and, but for this new bleaching process, it would scarcely have been possible for the cotton manufacture of Great Britain to have attained its present enormous extent, – it could not have competed in price with France and Germany. In the old process of bleaching, every piece must be exposed to the air and light during several weeks in the summer, and kept continually moist by manual labour. For this purpose, meadow land, eligibly situated, was essential. Now a single establishment near Glasgow bleaches 1400 pieces of cotton daily, throughout the year. What an enormous capital would be required to purchase land for this purpose! How greatly would it increase the cost of bleaching to pay interest upon this capital, or to hire so much land in England! This expense would scarcely have been felt in Germany. Besides the diminished expense, the cotton stuffs bleached with chlorine suffer less in the hands of skilful workmen than those bleached in the sun; and already the peasantry in some parts of Germany have adopted it, and find it advantageous.

Another use to which cheap muriatic acid is applied, is the manufacture of glue from bones. Bone contains from 30 to 36 per cent. of earthy matter – chiefly phosphate of lime, and the remainder is gelatine. When bones are digested in muriatic acid they become transparent and flexible like leather, the earthy matter is dissolved, and after the acid is all carefully washed away, pieces of glue of the same shape as the bones remain, which are soluble in hot water and adapted to all the purposes of ordinary glue, without further preparation.

Another important application of sulphuric acid may be adduced; namely, to the refining of silver and the separation of gold, which is always present in some proportion in native silver. Silver, as it is usually obtained from mines in Europe, contains in 16 ounces, 6 to 8 ounces of copper. When used by the silversmith, or in coining, 16 ounces must contain in Germany 13 ounces of silver, in England about 14 1/2. But this alloy is always made artificially by mixing pure silver with the due proportion of the copper; and for this purpose the silver must be obtained pure by the refiner. This he formerly effected by amalgamation, or by roasting it with lead; and the cost of this process was about 2l. for every hundredweight of silver. In the silver so prepared, about 1/1200 to 1/2000th part of gold remained; to effect the separation of this by nitrio-hydrochloric acid was more expensive than the value of the gold; it was therefore left in utensils, or circulated in coin, valueless. The copper, too, of the native silver was no use whatever. But the 1/1000th part of gold, being about one and a half per cent. of the value of the silver, now covers the cost of refining, and affords an adequate profit to the refiner; so that he effects the separation of the copper, and returns to his employer the whole amount of the pure silver, as well as the copper, without demanding any payment: he is amply remunerated by that minute portion of gold. The new process of refining is a most beautiful chemical operation: the granulated metal is boiled in concentrated sulphuric acid, which dissolves both the silver and the copper, leaving the gold nearly pure, in the form of a black powder. The solution is then placed in a leaden vessel containing metallic copper; this is gradually dissolved, and the silver precipitated in a pure metallic state. The sulphate of copper thus formed is also a valuable product, being employed in the manufacture of green and blue pigments.

Other immediate results of the economical production of sulphuric acid, are the general employment of phosphorus matches, and of stearine candles, that beautiful substitute for tallow and wax. Twenty-five years ago, the present prices and extensive applications of sulphuric and muriatic acids, of soda, phosphorus, would have been considered utterly impossible. Who is able to foresee what new and unthought-of chemical productions, ministering to the service and comforts of mankind, the next twenty-five years may produce?

After these remarks you will perceive that it is no exaggeration to say, we may fairly judge of the commercial prosperity of a country from the amount of sulphuric acid it consumes. Reflecting upon the important influence which the price of sulphur exercises upon the cost of production of bleached and printed cotton stuffs, soap, glass, and remembering that Great Britain supplies America, Spain, Portugal, and the East, with these, exchanging them for raw cotton, silk, wine, raisins, indigo, we can understand why the English Government should have resolved to resort to war with Naples, in order to abolish the sulphur monopoly, which the latter power attempted recently to establish. Nothing could be more opposed to the true interests of Sicily than such a monopoly; indeed, had it been maintained a few years, it is highly

probable that sulphur, the source of her wealth, would have been rendered perfectly valueless to her. Science and industry form a power to which it is dangerous to present impediments. It was not difficult to perceive that the issue would be the entire cessation of the exportation of sulphur from Sicily. In the short period the sulphur monopoly lasted, fifteen patents were taken out for methods to obtain back the sulphuric acid used in making soda. Admitting that these fifteen experiments were not perfectly successful, there can be no doubt it would ere long have been accomplished. But then, in gypsum, (sulphate of lime), and in heavy-spar, (sulphate of barytes), we possess mountains of sulphuric acid; in galena, (sulphate of lead), and in iron pyrites, we have no less abundance of sulphur. The problem is, how to separate the sulphuric acid, or the sulphur, from these native stores. Hundreds of thousands of pounds weight of sulphuric acid were prepared from iron pyrites, while the high price of sulphur consequent upon the

monopoly lasted. We should probably ere long have triumphed over all difficulties, and have separated it from gypsum. The impulse has been given, the possibility of the process proved, and it may happen in a few years that the inconsiderate financial speculation of Naples may deprive her of that lucrative commerce. In like manner Russia, by her prohibitory system, has lost much of her trade in tallow and potash. One country purchases only from absolute necessity from another, which excludes her own productions from her markets. Instead of the tallow and linseed oil of Russia, Great Britain now uses palm oil and cocoa-nut oil of other countries. Precisely analogous is the combination of workmen against their employers, which has led to the construction of many admirable machines for superseding manual labour. In commerce and industry every imprudence carries with it its own punishment; every oppression immediately and sensibly recoils upon the head of those from whom it emanates.

LETTER IV

My dear Sir,

One of the most influential causes of improvement in the social condition of mankind is that spirit of enterprise which induces men of capital to adopt and carry out suggestions for the improvement of machinery, the creation of new articles of commerce, or the cheaper production of those already in demand; and we cannot but admire the energy with which such men devote their talents, their time, and their wealth, to realise the benefits of the discoveries and inventions of science. For even when these are expended upon objects wholly incapable of realisation, – nay, even when the idea which first gave the impulse proves in the end to be altogether impracticable or absurd, immediate good to the community generally ensues; some useful and perhaps unlooked-for result flows directly, or springs ultimately, from exertions frustrated in their main design. Thus it is also in the pursuit of science. Theories lead to experiments and investigations; and he who investigates will scarcely ever fail of being rewarded by discoveries. It may be, indeed, the theory sought to be established is entirely unfounded in nature; but while searching in a right spirit for one thing, the inquirer may be rewarded by finding others far more valuable than those which he sought.

At the present moment, electromagnetism, as a moving power, is engaging great attention and study; wonders are expected from its application to this purpose. According to the sanguine expectations of many persons, it will shortly be employed to put into motion every kind of machinery, and amongst other things it will be applied to impel the carriages of railroads, and this at so small a cost, that expense will no longer be matter of consideration. England is to lose her superiority as a manufacturing country, inasmuch as her vast store of coals will no longer avail her as an economical source of motive power. “We,” say the German cultivators of this science, “have cheap zinc, and, how small a quantity of this metal is required to turn a lathe, and consequently to give motion to any kind of machinery!”

Such expectations may be very attractive, and yet they are altogether illusory! they will not bear the test of a few simple calculations; and these our friends have not troubled themselves to institute.

With a simple flame of spirits of wine, under a proper vessel containing boiling water, a small carriage of 200 to 300 pounds weight can be put into motion, or a weight of 80 to 100 pounds may be raised to a height of 20 feet. The same effects may be produced by dissolving zinc in dilute sulphuric acid in a certain apparatus. This is certainly an astonishing and highly interesting discovery; but the question to be determined is, which of the two processes is the least expensive?

In order to answer this question, and to judge correctly of the hopes entertained from this discovery, let me remind you of what chemists denominate "equivalents." These are certain unalterable ratios of effects which are proportionate to each other, and may therefore be expressed in numbers. Thus, if we require 8 pounds of oxygen to produce a certain effect, and we wish to employ chlorine for the same effect, we must employ neither more nor less than 35 1/2 pounds weight. In the same manner, 6 pounds weight of coal are equivalent to 32 pounds weight of zinc. The numbers representing chemical equivalents express very general ratios of effects, comprehending for all bodies all the actions they are capable of producing.

If zinc be combined in a certain manner with another metal, and submitted to the action of dilute sulphuric acid, it is dissolved in the form of an oxide; it is in fact burned at the expense of the oxygen contained in the fluid. A consequence of this action is the production of an electric current, which, if conducted through a wire, renders it magnetic. In thus effecting the solution of a pound weight, for example, of zinc, we obtain a definite amount of force adequate to raise a given weight one inch, and to keep it suspended; and the amount of weight it will be capable of suspending will be the greater the more rapidly the zinc is dissolved.

By alternately interrupting and renewing the contact of the zinc with the acid, and by very simple mechanical arrangements, we can give to the iron an upward and downward or a horizontal motion, thus producing the conditions essential to the motion of any machinery.

This moving force is produced by the oxidation of the zinc; and, setting aside the name given to the force in this case, we know that it can be produced in another manner. If we burn the zinc under the boiler of a steam-engine, consequently in the oxygen of the air instead of the galvanic pile, we should produce steam, and by it a certain amount of force. If we should assume, (which, however, is not proved,) that the quantity of force is unequal in these cases, – that, for instance, we had obtained double or triple the amount in the galvanic pile, or that in this mode of generating force less loss is sustained, – we must still recollect the equivalents of zinc and coal, and make these elements of our calculation. According to the experiments of Despretz, 6 pounds weight of zinc, in combining with oxygen, develops no more heat than 1 pound of coal; consequently, under equal conditions, we can produce six times the amount of force with a pound of coal as with a pound of zinc. It is therefore obvious that it would be more advantageous to employ coal instead of zinc, even if the latter produced four times as much force in a galvanic pile, as an equal weight of coal by its combustion under a boiler. Indeed it is highly probable, that if we burn under the boiler of a steam-engine the quantity of coal required for smelting the zinc from its ores, we shall produce far more force than the whole of the zinc so obtained could originate in any form of apparatus whatever.

Heat, electricity, and magnetism, have a similar relation to each other as the chemical equivalents of coal, zinc, and oxygen. By a certain measure of electricity we produce a corresponding proportion of heat or of magnetic power; we obtain that electricity by chemical affinity, which in one shape produces heat, in another electricity or magnetism. A certain amount of affinity produces an equivalent of electricity in the same manner as, on the other hand, we decompose equivalents of chemical compounds by a definite measure of electricity. The magnetic force of the pile is therefore limited to the extent of the chemical affinity, and in the case before us is obtained by the combination of the zinc and sulphuric acid. In the combustion of coal, the heat results from, and is measured by, the affinity of the oxygen of the atmosphere for that substance.

It is true that with a very small expense of zinc, we can make an iron wire a magnet capable of sustaining a thousand pounds weight of iron; let us not allow ourselves to be misled by this. Such a magnet could not raise a single pound weight of iron two inches, and therefore could not impart motion. The magnet acts like a rock, which while at rest presses with a weight of a thousand pounds upon a basis; it is like an inclosed lake, without an outlet and without a fall. But it may be said, we have, by mechanical arrangements, given it an outlet and a fall. True; and this must be regarded as a great triumph of mechanics; and I believe it is susceptible of further improvements, by which greater force may be obtained. But with every conceivable advantage of mechanism, no one will dispute that one pound of coal, under the boiler of a steam-engine, will give motion to a mass several hundred times greater than a pound of zinc in the galvanic pile.

Our experience of the employment of electromagnetism as a motory power is, however, too recent to enable us to foresee the ultimate results of contrivances to apply it; and, therefore, those who have devoted themselves to solve the problem of its application should not be discouraged, inasmuch as it would undoubtedly be a most important achievement to supersede the steam-engine, and thus escape the danger of railroads, even at double their expense.

Professor Weber of Gottingen has thrown out a suggestion, that if a contrivance could be devised to enable us to convert at will the wheels of the steam-carriage into magnets, we should be enabled to ascend and descend acclivities with great facility. This notion may ultimately be, to a certain extent, realised.

The employment of the galvanic pile as a motory power, however, must, like every other contrivance, depend upon the question of its relative economy: probably some time hence it may so far succeed as to be adopted in certain favourable localities; it may stand in the same relation to steam power as the manufacture of beet sugar bears to that of cane, or as the production of gas from oils and resins to that from mineral coal.

The history of beetroot sugar affords us an excellent illustration of the effect of prices upon commercial productions. This branch of industry seems at length, as to its processes, to be perfected. The most beautiful white sugar is now manufactured from the beetroot, in the place of the treacle-like sugar, having the taste of the root, which was first obtained; and instead of 3 or 4 per cent., the proportion obtained by Achard, double or even treble that amount is now produced. And notwithstanding the perfection of the manufacture, it is probable it will ere long be in most places entirely discontinued. In the years 1824 to 1827, the prices of agricultural produce were much lower than at present, while the price of sugar was the same. At that time one malter[1] of wheat was 10s., and one klafter[2] of wood 18s., and land was falling in price. Thus, food and fuel were cheap, and the demand for sugar unlimited; it was, therefore, advantageous to grow beetroot, and to dispose of the produce of land as sugar. All these circumstances are now different. A malter of wheat costs 18s.; a klafter of wood, 30s. to 36s. Wages have risen, but not in proportion, whilst the price of colonial sugar has fallen. Within the limits of the German commercial league, as, for instance, at Frankfort-on-the-Maine, a pound of the whitest and best loaf sugar is 7d.; the

---

1 Malter – a measure containing several bushels, but varying in different countries.]

2 Klafter – a cord, a stack, measuring six feet every way.

import duty is 31/d., or 30s. per cwt., leaving 31/d. as the price of the sugar. In the year 1827, then, one malter of wheat was equal to 40 lbs. weight of sugar, whilst at present that quantity of wheat is worth 70 lbs. of sugar. If indeed fuel were the same in price as formerly, and 70 lbs. of sugar could be obtained from the same quantity of the root as then yielded 40 lbs., it might still be advantageously produced; but the amount, if now obtained by the most approved methods of extraction, falls far short of this; and as fuel is double the price, and labour dearer, it follows that, at present, it is far more advantageous to cultivate wheat and to purchase sugar.

There are, however, other elements which must enter into our calculations; but these serve to confirm our conclusion that the manufacture of beetroot sugar as a commercial speculation must cease. The leaves and residue of the root, after the juice was expressed, were used as food for cattle, and their value naturally increased with the price of grain. By the process formerly pursued, 75 lbs. weight of juice were obtained from 100 lbs. of beetroot, and gave 5 lbs. of sugar. The method of Schutzenbach, which was eagerly adopted by the manufacturers, produced from the same quantity of root 8 lbs. of sugar; but it was attended with more expense to produce, and the loss of the residue as food for cattle. The increased expense in this process arises from the larger quantity of fuel required to evaporate the water; for instead of merely evaporating the juice, the dry residue is treated with water, and we require fuel sufficient to evaporate 106 lbs. of fluid instead of 75 lbs., and the residue is only fit for manure. The additional 3 lbs. of sugar are purchased at the expense of much fuel, and the loss of the residue as an article of food.

If the valley of the Rhine possessed mines of diamonds as rich as those of Golconda, Visiapoor, or the Brazils, they would probably not be worth the working: at those places the cost of extraction is 28s. to 30s. the carat. With us it amounts to three or four times as much – to more, in fact, than diamonds are worth in the market. The sand of the Rhine contains gold; and in the Grand Duchy of Baden many persons are occupied in gold-washing when wages are low; but as soon as they rise, this employment ceases. The manufacture of sugar from beetroot, in the like manner, twelve to fourteen years ago offered advantages which are now lost: instead, therefore, of maintaining it at a great sacrifice, it would be more reasonable, more in accordance with true natural economy, to cultivate other and more valuable productions, and with them purchase sugar. Not only would the state be the gainer, but every member of the community. This argument does not apply, perhaps, to France and Bohemia, where the prices of fuel and of colonial sugar are very different to those in Germany.

The manufacture of gas for lighting, from coal, resin, and oils, stands with us on the same barren ground.

The price of the materials from which gas is manufactured in England bears a direct proportion to the price of corn: there the cost of tallow and oil is twice as great as in Germany, but iron and coal are two-thirds cheaper; and even in England the manufacture of gas is only advantageous when the other products of the distillation of coal, the coke, can be sold.

It would certainly be esteemed one of the greatest discoveries of the age if any one could succeed in condensing coal gas into a white, dry, solid, odourless substance, portable, and capable of being placed upon a candlestick, or burned in a lamp. Wax, tallow, and oil, are combustible gases in a solid or fluid form, which offer many advantages for lighting, not possessed by gas: they furnish, in well-constructed lamps, as much light, without requiring the expensive apparatus necessary for the combustion of gas, and they are generally more economical. In large towns, or such establishments as hotels, where coke is in demand, and where losses in stolen tallow or oil must be considered, together with the labour of snuffing candles and cleaning lamps, the higher price of gas is compensated. In places where gas can be manufactured from resin, oil of turpentine, and other cheap oils, as at Frankfort, this is advantageous so long as it is pursued on small scale only. If large towns were lighted in the same manner, the materials would rise in price: the whole amount at present produced would scarcely suffice for two such towns as Berlin and Munich. But no just calculation can be made from the present prices of turpentine, resin, which are not produced upon any large scale.

LETTER V

My dear Sir,

Until very recently it was supposed that the physical qualities of bodies, i.e. hardness, colour, density, transparency, and still more their chemical properties, must depend upon the nature of their elements, or upon their composition. It was tacitly received as a principle, that two bodies containing the same elements in the same proportion, must of necessity possess the same properties. We could not imagine an exact identity of composition giving rise to two bodies entirely different in their sensible appearance and chemical relations. The most ingenious philosophers entertained the opinion that chemical combination is an Interpenetration of the particles of different kinds of matter, and that all matter is susceptible of infinite division. This has proved to be altogether a mistake. If matter were infinitely divisible in this sense, its particles must be imponderable, and a million of such molecules could not weigh more than an infinitely small one. But the particles of that imponderable matter, which, striking upon the retina, give us the sensation of light, are not in a mathematical sense infinitely small.

Interpenetration of elements in the production of a chemical compound, supposes two distinct bodies, A and B, to occupy one and the same space at the same time. If this were so, different properties could not consist with an equal and identical composition.

That hypothesis, however, has shared the fate of innumerable imaginative explanations of natural phenomena, in which our predecessors indulged. They have now no advocate. The force of truth, dependent upon observation, is irresistible. A great many substances have been discovered amongst organic bodies, composed of the same elements in the same relative proportions, and yet exhibiting physical and chemical properties perfectly distinct one from another. To such substances the term Isomeric (from 1/ao1/ equal and aei1/o1/ part) is applied. A great class of bodies, known as the volatile oils, oil of turpentine, essence of lemons, oil of balsam of copaiba, oil of rosemary, oil of juniper, and many others, differing widely from each other in their odour, in their medicinal effects, in their boiling point, in their specific gravity, are exactly identical in composition, – they contain the same elements, carbon and hydrogen, in the same proportions.

How admirably simple does the chemistry of organic nature present itself to us from this point of view! An extraordinary variety of compound bodies produced with equal weights of two elements! and how wide their dissimilarity! The crystallised part of the oil of roses, the delicious fragrance of which is so well known, a solid at ordinary temperatures, although readily volatile, is a compound body containing exactly the same elements, and in the same proportions, as the gas we employ for lighting our streets; and, in short, the same elements, in the same relative quantities, are found in a dozen other compounds, all differing essentially in their physical and chemical properties.

These remarkable truths, so highly important in their applications, were not received and admitted as sufficiently established, without abundant proofs. Many examples have long been known where the analysis of two different bodies gave the same composition; but such cases were regarded as doubtful: at any rate, they were isolated observations, homeless in the realms of science: until, at length, examples were discovered of two or more bodies whose absolute identity of composition, with totally distinct properties, could be demonstrated in a more obvious and conclusive manner than by mere analysis; that is, they can be converted and reconverted into each other without addition and without subtraction.

In cyanuric acid, hydrated cyanic acid, and cyamelide, we have three such isomeric compounds.

Cyanuric acid is crystalline, soluble in water, and capable of forming salts with metallic oxides.

Hydrated cyanic acid is a volatile and highly blistering fluid, which cannot be brought into contact with water without being instantaneously decomposed.

Cyamelide is a white substance very like porcelain, absolutely insoluble in water.

Now if we place the first, – cyanuric acid, – in a vessel hermetically sealed, and apply a high degree of heat, it is converted by its influence into hydrated cyanic acid; and, then, if this is kept for some time at the common temperature, it passes into cyamelide, no other element being present. And, again inversely, cyamelide can be converted into cyanuric acid and hydrated cyanic acid.

We have three other bodies which pass through similar changes, in aldehyde, metaldehyde, and etaldehyde; and, again two, in urea and cyanuret of ammonia. Further, 100 parts of aldehyde hydrated butyric acid and acetic ether contain the same elements in the same proportion. Thus one substance may be converted into another without addition or subtraction, and without the participation of any foreign bodies in the change.

The doctrine that matter is not infinitely divisible, but on the contrary, consists of atoms incapable of further division, alone furnishes us with a satisfactory explanation of these phenomena. In chemical combinations, the ultimate atoms of bodies do not penetrate each other, they are only arranged side by side in a certain order, and the properties of the compound depend entirely upon this order. If they are made to change their place – their mode of arrangement – by an impulse from without, they combine again in a different manner, and another compound is formed with totally different properties. We may suppose that one atom combines with one atom of another element to form a compound atom, while in other bodies two and two, four and four, eight and eight, are united; so that in all such compounds the amount per cent. of the elements is absolutely equal; and yet their physical and chemical properties must be totally different, the constitution of each atom being peculiar, in one body consisting of two, in another of four, in a third of eight, and in a fourth of sixteen simple atoms.

The discovery of these facts immediately led to many most beautiful and interesting results; they furnished us with a satisfactory explanation of observations which were before veiled in mystery, – a key to many of Nature's most curious recesses.

Again; solid bodies, whether simple or compound, are capable of existing in two states, which are known by the terms amorphous and crystalline.

When matter is passing from a gaseous or liquid state slowly into a solid, an incessant motion is observed, as if the molecules were minute magnets; they are seen to repel each other in one direction, and to attract and cohere together in another, and in the end become arranged into a regular form, which under equal circumstances is always the same for any given kind of matter; that is, crystals are formed.

Time and freedom of motion for the particles of bodies are necessary to the formation of crystals. If we force a fluid or a gas to become suddenly solid, leaving no time for its particles to arrange themselves, and cohere in that direction in which the cohesive attraction is strongest, no crystals will be formed, but the resulting solid will have a different colour, a different degree of hardness and cohesion, and will refract light differently; in one word, will be amorphous. Thus we have cinnabar as a red and a jet-black substance; sulphur a fixed and brittle body, and soft, semitransparent, and ductile; glass as a milk-white opaque substance, so hard that it strikes fire with steel, and in its ordinary and well-known state. These dissimilar states and properties of the same body are occasioned in one case by a regular, in the other by an irregular, arrangement of its atoms; one is crystalline, the other amorphous.

Applying these facts to natural productions, we have reason to believe that clay-slate, and many kinds of greywacke, are amorphous feldspar, as transition limestone is amorphous marble, basalt and lava mixtures of amorphous zeolite and augite. Anything that influences the cohesion, must also in a certain degree alter the properties of bodies. Carbonate of lime, if crystallised at ordinary temperatures, possesses the crystalline form, hardness, and refracting power of common spar; if crystallised at a higher temperature, it has the form and properties of arragonite.

Finally, Isomorphism, or the equality of form of many chemical compounds having a different composition, tends to prove that matter consists of atoms the mere arrangement of which produces all the properties of bodies. But when we find that a different arrangement of the same elements gives rise to various physical and chemical properties, and a similar arrangement of different elements produces properties very much the same, may we not inquire whether some of those bodies which we regard as elements may not be merely modifications of the same substance? – whether they are not the same matter in a different state of arrangement? We know in fact the existence of iron in two states, so dissimilar, that in the one, it is to the electric chain like platinum, and in the other it is like zinc; so that powerful galvanic machines have been constructed of this one metal.

Among the elements are several instances of remarkable similarity of properties. Thus there is a strong resemblance between platinum and iridium; bromine and iodine; iron, manganese, and magnesium; cobalt and nickel; phosphorus and arsenic; but this resemblance consists mainly in their forming isomorphous compounds in which these elements exist in the same relative proportion. These compounds are similar, because the atoms of which they are composed are arranged in the same manner. The converse of this is also true: nitrate of strontia becomes quite dissimilar to its common state if a certain proportion of water is taken into its composition.

If we suppose selenium to be merely modified sulphur, and phosphorus modified arsenic, how does it happen, we must inquire, that sulphuric acid and selenic acid, phosphoric and arsenic acid, respectively form compounds which it is impossible to distinguish by their form and solubility? Were these merely isomeric, they ought to exhibit properties quite dissimilar!

We have not, I believe, at present the remotest ground to suppose that any one of those substances which chemists regard as elements can be converted into another. Such a conversion, indeed, would presuppose that the element was composed of two or more ingredients, and was in fact not an element; and until the decomposition of these bodies is accomplished, and their constituents discovered, all pretensions to such conversions deserve no notice.

Dr. Brown of Edinburgh thought he had converted iron into rhodium, and carbon or paracyanogen into silicon. His paper upon this subject was published in the Transactions of the Royal Society of Edinburgh, and contained internal evidence, without a repetition of his

experiments, that he was totally unacquainted with the principles of chemical analysis. But his experiments have been carefully repeated by qualified persons, and they have completely proved his ignorance: his rhodium is iron, and his silicon an impure incombustible coal.

LETTER VI

My dear Sir,

One of the most remarkable effects of the recent progress of science is the alliance of chemistry with physiology, by which a new and unexpected light has been thrown upon the vital processes of plants and animals. We have now no longer any difficulty in understanding the different actions of aliments, poisons, and remedial agents – we have a clear conception of the causes of hunger, of the exact nature of death; and we are not, as formerly, obliged to content ourselves with a mere description of their symptoms. It is now ascertained with positive certainty, that all the substances which constitute the food of man must be divided into two great classes, one of which serves for the nutrition and reproduction of the animal body, whilst the other ministers to quite different purposes. Thus starch, gum, sugar, beer, wine, spirits, furnish no element capable of entering into the composition of blood, muscular fibre, or any part which is the seat of the vital principle. It must surely be universally interesting to trace the great change our views have undergone upon these subjects, as well as to become acquainted with the researches from which our present knowledge is derived.

The primary conditions of the maintenance of animal life, are a constant supply of certain matters, animal food, and of oxygen, in the shape of atmospheric air. During every moment of life, oxygen is absorbed from the atmosphere in the organs of respiration, and the act of breathing cannot cease while life continues.

The observations of physiologists have demonstrated that the body of an adult man supplied abundantly with food, neither increases nor diminishes in weight during twenty-four hours, and yet the quantity of oxygen absorbed into his system, in that period, is very considerable. According to the experiments of Lavoisier, an adult man takes into his system from the atmosphere, in one year, no less than 746 pounds weight of oxygen; the calculations of Menzies make the quantity amount even to 837 pounds; but we find his weight at the end of the year either exactly the same or different one way or the other by at most a few pounds. What, it may be asked, has become of the enormous amount of oxygen thus introduced into the human system in the course of one year? We can answer this question satisfactorily. No part of the oxygen remains in the body, but is given out again, combined with carbon and hydrogen. The carbon and hydrogen of certain parts of the animal body combine with the oxygen introduced through the lungs and skin, and pass off in the forms of carbonic acid and vapour of water. At every expiration and every moment of life, a certain amount of its elements are separated from the animal organism, having entered into combination with the oxygen of the atmosphere.

In order to obtain a basis for the approximate calculation, we may assume, with Lavoisier and Seguin, that an adult man absorbs into his system 32 1/2 ounces of oxygen daily, – that is, 46,037 cubic inches = 15,661 grains, French weight; and further, that the weight of the whole mass of his blood is 24 pounds, of which 80 per cent. is water. Now, from the known composition of the blood, we know that in order to convert its whole amount of carbon and hydrogen into carbonic acid and water, 64.102 grains of oxygen are required. This quantity will be taken into the system in four days and five hours. Whether the oxygen enters into combination directly with the elements of the blood, or with the carbon and hydrogen of other parts of the body, it follows inevitably – the weight of the body remaining unchanged and in a normal condition – that as much of these elements as will suffice to supply 24 pounds of blood, must be taken into the system in four days and five hours; and this necessary amount is furnished by the food.

We have not, however, remained satisfied with mere approximation: we have determined accurately, in certain cases, the quantity of carbon taken daily in the food, and of that which passes out of the body in the faeces and urine combined – that is, uncombined with oxygen; and from these investigations it appears that an adult man taking moderate exercise consumes 13.9 ounces of carbon, which pass off through the skin and lungs as carbonic acid gas.[3]

---

3 This account is deduced from observations made upon the average daily consumption of about 30 soldiers in barracks. The food of these men, consisting of meat, bread, potatoes, lentils, peas, beans, butter, salt, pepper, was accurately weighed during a month, and each article subjected to ultimate analysis. Of the quantity of food, beer, and spirits, taken by the men when out of barracks, we have a close approximation from the report of the sergeant; and from the

It requires 37 ounces of oxygen to convert 13 9/10 of carbon into carbonic acid. Again; according to the analysis of Boussingault, (Annales de Chim. et de Phys., lxx. i. p.136), a horse consumes 79 1/10 ounces of carbon in twenty-four hours, a milch cow 70 3/4 ounces; so that the horse requires 13 pounds 3 1/2 ounces, and the cow 11 pounds 10 3/4 ounces of oxygen.[4]

As no part of the oxygen taken into the system of an animal is given off in any other form than combined with carbon or hydrogen, and as in a normal condition, or state of health, the carbon and hydrogen so given off are replaced by those elements in the food,

---

weight and analysis of the faeces and urine, it appears that the carbon which passes off through these channels may be considered equivalent to the amount taken in that portion of the food, and of sourcrout, which was not included in the estimate.

4 17.5 ounces = 0.5 kilogramme.

it is evident that the amount of nourishment required by an animal for its support must be in a direct ratio with the quantity of oxygen taken in to its system. Two animals which in equal times take up by means of the lungs and skin unequal quantities of oxygen, consume an amount of food unequal in the same ratio. The consumption of oxygen in a given time may be expressed by the number of respirations; it is, therefore, obvious that in the same animal the quantity of nourishment required must vary with the force and number of respirations. A child breathes quicker than an adult, and, consequently, requires food more frequently and proportionably in larger quantity, and bears hunger less easily. A bird deprived of food dies on the third day, while a serpent, confined under a bell, respires so slowly that the quantity of carbonic acid generated in an hour can scarcely be observed, and it will live three months, or longer, without food. The number of respirations is fewer in a state of rest than during labour or exercise: the quantity of food necessary in both cases must be in the same ratio. An excess of food, a want of a due amount of respired oxygen, or of exercise, as also great exercise (which obliges us to take an increased supply of food), together with weak organs of digestion, are incompatible with health

But the quantity of oxygen received by an animal through the lungs not only depends upon the number of respirations, but also upon the temperature of the respired air. The size of the thorax of an animal is unchangeable; we may therefore regard the volume of air which enters at every inspiration as uniform. But its weight, and consequently the amount of oxygen it contains, is not constant. Air is expanded by heat, and contracted by cold – an equal volume of hot and cold air contains, therefore, an unequal amount of oxygen. In summer atmospheric air contains water in the form of vapour, it is nearly deprived of it in winter; the volume of oxygen in the same volume of air is smaller in summer than in winter. In summer and winter, at the pole and at the equator, we inspire an equal volume of air; the cold air is warmed during respiration and acquires the temperature of the body. In order, therefore, to introduce into the lungs a given amount of oxygen, less expenditure of force is necessary in winter than in summer, and for the same expenditure of force more oxygen is inspired in winter. It is also obvious that in an equal number of respirations we consume more oxygen at the level of the sea than on a mountain.

The oxygen taken into the system is given out again in the same form, both in summer and winter: we expire more carbon at a low than at a high temperature, and require more or less carbon in our food in the same proportion; and, consequently, more is respired in Sweden than in Sicily, and in our own country and eighth more in winter than in summer. Even if an equal weight of food is consumed in hot and cold climates, Infinite Wisdom has ordained that very unequal proportions of carbon shall be taken in it. The food prepared for the inhabitants of southern climes does not contain in a fresh state more than 12 per cent. of carbon, while the blubber and train oil which feed the inhabitants of Polar regions contain 66 to 80 per cent. of that element.

From the same cause it is comparatively easy to be temperate in warm climates, or to bear hunger for a long time under the equator; but cold and hunger united very soon produce exhaustion.

The oxygen of the atmosphere received into the blood in the lungs, and circulated throughout every part of the animal body, acting upon the elements of the food, is the source of animal heat.

LETTER VII

My dear Sir,

The source of animal heat, its laws, and the influence it exerts upon the functions of the animal body, constitute a curious and highly interesting subject, to which I would now direct your attention.

All living creatures, whose existence depends upon the absorption of oxygen, possess within themselves a source of heat, independent of surrounding objects.

This general truth applies to all animals, and extends to the seed of plants in the act of germination, to flower-buds when developing, and fruits during their maturation.

In the animal body, heat is produced only in those parts to which arterial blood, and with it the oxygen absorbed in respiration, is conveyed. Hair, wool, and feathers, receive no arterial blood, and, therefore, in them no heat is developed. The combination of a combustible substance with oxygen is, under all circumstances, the only source of animal heat. In whatever way carbon may combine with oxygen, the act of combination is accompanied by the disengagement of heat. It is indifferent whether this combination takes place rapidly or slowly, at a high or at a low temperature: the amount of heat liberated is a constant quantity.

The carbon of the food, being converted into carbonic acid within the body, must give out exactly as much heat as if it had been directly burnt in oxygen gas or in common air; the only difference is, the production of the heat is diffused over unequal times. In oxygen gas the combustion of carbon is rapid and the heat intense; in atmospheric air it burns slower and for a longer time, the temperature being lower; in the animal body the combination is still more gradual, and the heat is lower in proportion.

It is obvious that the amount of heat liberated must increase or diminish with the quantity of oxygen introduced in equal times by respiration. Those animals, therefore, which respire frequently, and consequently consume much oxygen, possess a higher temperature than others, which, with a body of equal size to be heated, take into the system less oxygen. The temperature of a child (102 deg) is higher than that of an adult (99 1/2 deg). That of birds (104 deg to 105.4 deg) is higher than that of quadrupeds (98 1/2 deg to 100.4 deg) or than that of fishes or amphibia, whose proper temperature is from 2.7 to 3.6 deg higher than that of the medium in which they live. All animals, strictly speaking, are warm-blooded; but in those only which possess lungs is the temperature of the body quite independent of the surrounding medium.

The most trustworthy observations prove that in all climates, in the temperate zones as well as at the equator or the poles, the temperature of the body in man, and in what are commonly called warm-blooded animals, is invariably the same; yet how different are the circumstances under which they live!

The animal body is a heated mass, which bears the same relation to surrounding objects as any other heated mass. It receives heat when the surrounding objects are hotter, it loses heat when they are colder, than itself.

We know that the rapidity of cooling increases with the difference between the temperature of the heated body and that of the surrounding medium; that is, the colder the surrounding medium the shorter the time required for the cooling of the heated body.

How unequal, then, must be the loss of heat in a man at Palermo, where the external temperature is nearly equal to that of the body, and in the polar regions, where the external temperature is from 70 deg to 90 deg lower!

Yet, notwithstanding this extremely unequal loss of heat, experience has shown that the blood of the inhabitant of the arctic circle has a temperature as high as that of the native of the south, who lives in so different a medium.

This fact, when its true significance is perceived, proves that the heat given off to the surrounding medium is restored within the body with great rapidity. This compensation must consequently take place more rapidly in winter than in summer, at the pole than at the equator.

Now, in different climates the quantity of oxygen introduced into the system by respiration, as has been already shown, varies according to the temperature of the external air; the quantity of inspired oxygen increases with the loss of heat by external cooling, and the quantity of carbon or hydrogen necessary to combine with this oxygen must be increased in the same ratio.

It is evident that the supply of the heat lost by cooling is effected by the mutual action of the elements of the food and the inspired oxygen, which combine together. To make use of a familiar, but not on that account a less just illustration, the animal body acts, in this respect, as a furnace, which we supply with fuel. It signifies nothing what intermediate forms food may assume, what changes it may undergo in the body; the last change is uniformly the conversion of its carbon into carbonic acid, and of its hydrogen into water. The unassimilated nitrogen of the food, along with the unburned or unoxidised carbon, is expelled in the urine or in the solid excrements. In order to keep up in the furnace a constant temperature, we must vary the supply of fuel according to the external temperature, that is, according to the supply of oxygen.

In the animal body the food is the fuel; with a proper supply of oxygen we obtain the heat given out during its oxidation or combustion. In winter, when we take exercise in a cold atmosphere, and when consequently the amount of inspired oxygen increases, the necessity for food containing carbon and hydrogen increases in the same ratio; and by gratifying the appetite thus excited, we obtain the most efficient protection against the most piercing cold. A starving man is soon frozen to death. The animals of prey in the arctic regions, as every one knows, far exceed in voracity those of the torrid zone.

In cold and temperate climates, the air, which incessantly strives to consume the body, urges man to laborious efforts in order to furnish the means of resistance to its action, while, in hot climates, the necessity of labour to provide food is far less urgent.

Our clothing is merely an equivalent for a certain amount of food. The more warmly we are clothed the less urgent becomes the appetite for food, because the loss of heat by cooling, and consequently the amount of heat to be supplied by the food, is diminished.

If we were to go naked, like certain savage tribes, or if in hunting or fishing we were exposed to the same degree of cold as the Samoyedes, we should be able with ease to consume 10 lbs. of flesh, and perhaps a dozen of tallow candles into the bargain, daily, as warmly clad travellers have related with astonishment of these people. We should then also be able to take the same quantity of brandy or train oil without bad effects, because the carbon and hydrogen of these substances would only suffice to keep up the equilibrium between the external temperature and that of our bodies.

According to the preceding expositions, the quantity of food is regulated by the number of respirations, by the temperature of the air, and by the amount of heat given off to the surrounding medium.

No isolated fact, apparently opposed to this statement, can affect the truth of this natural law. Without temporary or permanent injury to health, the Neapolitan cannot take more carbon and hydrogen in the shape of food than he expires as carbonic acid and water; and the Esquimaux cannot expire more carbon and hydrogen than he takes in the system as food, unless in a state of disease or of starvation. Let us examine these states a little more closely.

The Englishman in Jamaica perceives with regret the disappearance of his appetite, previously a source of frequently recurring enjoyment; and he succeeds, by the use of cayenne pepper, and the most powerful stimulants, in enabling himself to take as much food as he was accustomed to eat at home. But the whole of the carbon thus introduced into the system is not consumed; the temperature of the air is too high, and the oppressive heat does not allow him to increase the number of respirations by active exercise, and thus to proportion the waste to the amount of food taken; disease of some kind, therefore, ensues.

On the other hand, England sends her sick to southern regions, where the amount of the oxygen inspired is diminished in a very large proportion. Those whose diseased digestive organs have in a greater or less degree lost the power of bringing the food into the state best adapted for oxidation, and therefore are less able to resist the oxidising influence of the atmosphere of their native climate, obtain a great improvement in health. The diseased organs of digestion have power to place the diminished amount of food in equilibrium with the inspired oxygen, in the mild climate; whilst in a colder region the organs of respiration themselves would have been consumed in furnishing the necessary resistance to the action of the atmospheric oxygen.

In our climate, hepatic diseases, or those arising from excess of carbon, prevail in summer; in winter, pulmonary diseases, or those arising from excess of oxygen, are more frequent.

The cooling of the body, by whatever cause it may be produced, increases the amount of food necessary. The mere exposure to the open air, in a carriage or on the deck of a ship, by increasing radiation and vaporisation, increases the loss of heat, and compels us to eat more than usual. The same is true of those who are accustomed to drink large quantities of cold water, which is given off at the temperature of the body, 98 1/2 deg. It increases the appetite, and persons of weak constitution find it necessary, by continued exercise, to supply to the system the oxygen required to restore the heat abstracted by the cold water. Loud and long continued speaking, the crying of infants, moist air, all exert a decided and appreciable influence on the amount of food which is taken.

We have assumed that carbon and hydrogen especially, by combining with oxygen, serve to produce animal heat. In fact, observation proves that the hydrogen of the food plays a no less important part than the carbon.

The whole process of respiration appears most clearly developed, when we consider the state of a man, or other animal, totally deprived of food.

The first effect of starvation is the disappearance of fat, and this fat cannot be traced either in the urine or in the scanty faeces. Its carbon and hydrogen have been given off through the skin and lungs in the form of oxidised products; it is obvious that they have served to support respiration.

In the case of a starving man, 32 1/2 oz. of oxygen enter the system daily, and are given out again in combination with a part of his body. Currie mentions the case of an individual who was unable to swallow, and whose body lost 100 lbs. in weight during a month; and, according to Martell (Trans. Linn. Soc., vol. xi. p.411), a fat pig, overwhelmed in a slip of earth, lived 160 days without food, and was found to have diminished in weight, in that time, more than 120 lbs. The whole history of hybernating animals, and the well-established facts of the periodical accumulation, in various animals, of fat, which, at other periods, entirely disappears, prove that the oxygen, in the respiratory process, consumes, without exception, all such substances as are capable of entering into combination with it. It combines with whatever is presented to it; and the deficiency of hydrogen is the only reason why carbonic acid is the chief product; for, at the temperature of the body, the affinity of hydrogen for oxygen far surpasses that of carbon for the same element.

We know, in fact, that the graminivora expire a volume of carbonic acid equal to that of the oxygen inspired, while the carnivora, the only class of animals whose food contains fat, inspire more oxygen than is equal in volume to the carbonic acid expired. Exact experiments have shown, that in many cases only half the volume of oxygen is expired in the form of carbonic acid. These observations cannot be gainsaid, and are far more convincing than those arbitrary and artificially produced phenomena, sometimes called experiments; experiments which, made as too often they are, without regard to the necessary and natural conditions, possess no value, and may be entirely dispensed with; especially when, as in the present case, Nature affords the opportunity for observation, and when we make a rational use of that opportunity.

In the progress of starvation, however, it is not only the fat which disappears, but also, by degrees all such of the solids as are capable of being dissolved. In the wasted bodies of those who have suffered starvation, the muscles are shrunk and unnaturally soft, and have lost their contractibility; all those parts of the body which were capable of entering into the state of motion have served to protect the remainder of the frame from the destructive influence of the atmosphere. Towards the end, the particles of the brain begin to undergo the process of oxidation, and delirium, mania, and death close the scene; that is to say, all resistance to the oxidising power of the atmospheric oxygen ceases, and the chemical process of eremacausis, or decay, commences, in which every part of the body, the bones excepted, enters into combination with oxygen.

The time which is required to cause death by starvation depends on the amount of fat in the body, on the degree of exercise, as in labour or exertion of any kind, on the temperature of the air, and finally, on the presence or absence of water. Through the skin and lungs there escapes a certain quantity of water, and as the presence of water is essential to the continuance of the vital motions, its dissipation hastens death. Cases have occurred, in which a full supply of water being accessible to the sufferer, death has not occurred till after the lapse of twenty days. In one case, life was sustained in this way for the period of sixty days.

In all chronic diseases death is produced by the same cause, namely, the chemical action of the atmosphere. When those substances are wanting, whose function in the organism is to support the process of respiration, when the diseased organs are incapable of performing their proper function of producing these substances, when they have lost the power of transforming the food into that shape in which it may, by entering into combination with the oxygen of the air, protect the system from its influence, then, the substance of the organs themselves, the fat of the body, the substance of the muscles, the nerves, and the brain, are unavoidably consumed.

The true cause of death in these cases is the respiratory process, that is, the action of the atmosphere.

A deficiency of food, and a want of power to convert the food into a part of the organism, are both, equally, a want of resistance; and this is the negative cause of the cessation of the vital process. The flame is extinguished, because the oil is consumed; and it is the oxygen of the air which has consumed it.

In many diseases substances are produced which are incapable of assimilation. By the mere deprivation of food, these substances are removed from the body without leaving a trace behind; their elements have entered into combination with the oxygen of the air.

From the first moment that the function of the lungs or of the skin is interrupted or disturbed, compounds, rich in carbon, appear in the urine, which acquires a brown colour. Over the whole surface of the body oxygen is absorbed, and combines with all the substances which offer no resistance to it. In those parts of the body where the access of oxygen is impeded; for example, in the armpits, or in the soles of the feet, peculiar compounds are given out, recognisable by their appearance, or by their odour. These compounds contain much carbon.

Respiration is the falling weight – the bent spring, which keeps the clock in motion; the inspirations and expirations are the strokes of the pendulum which regulate it. In our ordinary timepieces, we know with mathematical accuracy the effect produced on their rate of going, by changes in the length of the pendulum, or in the external temperature. Few, however, have a clear conception of the influence of air and temperature on the health

of the human body; and yet the research into the conditions necessary to keep it in the normal state is not more difficult than in the case of a clock.

LETTER VIII

My dear Sir,

Having attempted in my last letter to explain to you the simple and admirable office subserved by the oxygen of the atmosphere in its combination with carbon in the animal body, I will now proceed to present you with some remarks upon those materials which sustain its mechanisms in motion, and keep up their various functions, – namely, the Aliments.

If the increase in mass in an animal body, the development and reproduction of its organs depend upon the blood, then those substances only which are capable of being converted into blood can be properly regarded as nourishment. In order then to ascertain what parts of our food are nutritious, we must compare the composition of the blood with the composition of the various articles taken as food.

Two substances require especial consideration as the chief ingredients of the blood; one of these separates immediately from the blood when it is withdrawn from the circulation.

It is well known that in this case blood coagulates, and separates into a yellowish liquid, the serum of the blood, and a gelatinous mass, which adheres to a rod or stick in soft, elastic fibres, when coagulating blood is briskly stirred. This is the fibrine of the blood, which is identical in all its properties with muscular fibre, when the latter is purified from all foreign matters.

The second principal ingredient of the blood is contained in the serum, and gives to this liquid all the properties of the white of eggs, with which it is indeed identical. When heated, it coagulates into a white elastic mass, and the coagulating substance is called albumen.

Fibrine and albumen, the chief ingredients of blood, contain, in all, seven chemical elements, among which nitrogen, phosphorus, and sulphur are found. They contain also the earth of bones. The serum retains in solution sea salt and other salts of potash and soda, in which the acids are carbonic, phosphoric, and sulphuric acids. The globules of the blood contain fibrine and albumen, along with a red colouring matter, in which iron is a constant element. Besides these, the blood contains certain fatty bodies in small quantity, which differ from ordinary fats in several of their properties.

Chemical analysis has led to the remarkable result, that fibrine and albumen contain the same organic elements united in the same proportion, – i.e., that they are isomeric, their chemical composition – the proportion of their ultimate elements – being identical. But the difference of their external properties shows that the particles of which they are composed are arranged in a different order. (See Letter V).

This conclusion has lately been beautifully confirmed by a distinguished physiologist (Denis), who has succeeded in converting fibrine into albumen, that is, in giving it the solubility, and coagulability by heat, which characterise the white of egg.

Fibrine and albumen, besides having the same composition, agree also in this, that both dissolve in concentrated muriatic acid, yielding a solution of an intense purple colour. This solution, whether made with fibrine or albumen, has the very same reactions with all substances yet tried.

Both albumen and fibrine, in the process of nutrition, are capable of being converted into muscular fibre, and muscular fibre is capable of being reconverted into blood. These facts have long been established by physiologists, and chemistry has merely proved that these metamorphoses can be accomplished under the influence of a certain force, without the aid of a third substance, or of its elements, and without the addition of any foreign element, or the separation of any element previously present in these substances.

If we now compare the composition of all organised parts with that of fibrine and albumen, the following relations present themselves:–

All parts of the animal body which have a decided shape, which form parts of organs, contain nitrogen. No part of an organ which possesses motion and life is destitute of nitrogen; all of them contain likewise carbon and the elements of water; the latter, however, in no case in the proportion to form water.

The chief ingredients of the blood contain nearly 17 per cent. of nitrogen, and from numerous analyses it appears that no part of an organ contains less than 17 per cent. of nitrogen.

The most convincing experiments and observations have proved that the animal body is absolutely incapable of producing an elementary body, such as carbon or nitrogen, out of substances which do not contain it; and it obviously follows, that all kinds of food fit for the production either of blood, or of cellular tissue, membranes, skin, hair, muscular fibre, must contain a certain amount of nitrogen, because that element is essential to the composition of the above-named organs; because the organs cannot create it from the other elements presented to them; and, finally, because no nitrogen is absorbed from the atmosphere in the vital process.

The substance of the brain and nerves contains a large quantity of albumen, and, in addition to this, two peculiar fatty acids, distinguished from other fats by containing phosphorus (phosphoric acid?). One of these contains nitrogen (Fremy).

Finally, water and common fat are those ingredients of the body which are destitute of nitrogen. Both are amorphous or unorganised, and only so far take part in the vital process as that their presence is required for the due performance of the vital functions. The inorganic constituents of the body are, iron, lime, magnesia, common salt, and the alkalies.

The nutritive process is seen in its simplest form in carnivorous animals. This class of animals lives on the blood and flesh of the graminivora; but this blood and flesh are, in all their properties, identical with their own. Neither chemical nor physiological differences can be discovered.

The nutriment of carnivorous animals is derived originally from blood; in their stomach it becomes dissolved, and capable of reaching all other parts of the body; in its passage it is again converted into blood, and from this blood are reproduced all those parts of their organisation which have undergone change or metamorphosis.

With the exception of hoofs, hair, feathers, and the earth of bones, every part of the food of carnivorous animals is capable of assimilation.

In a chemical sense, therefore, it may be said that a carnivorous animal, in supporting the vital process, consumes itself. That which serves for its nutrition is identical with those parts of its organisation which are to be renewed.

The process of nutrition in graminivorous animals appears at first sight altogether different. Their digestive organs are less simple, and their food consists of vegetables, the great mass of which contains but little nitrogen.

From what substances, it may be asked, is the blood formed, by means of which of their organs are developed? This question may be answered with certainty.

Chemical researches have shown, that all such parts of vegetables as can afford nutriment to animals contain certain constituents which are rich in nitrogen; and the most ordinary experience proves that animals require for their support and nutrition less of these parts of plants in proportion as they abound in the nitrogenised constituents. Animals cannot be fed on matters destitute of these nitrogenised constituents.

These important products of vegetation are especially abundant in the seeds of the different kinds of grain, and of peas, beans, and lentils; in the roots and the juices of what are commonly called vegetables. They exist, however, in all plants, without exception, and in every part of plants in larger or smaller quantity.

These nitrogenised forms of nutriment in the vegetable kingdom may be reduced to three substances, which are easily distinguished by their external characters. Two of them are soluble in water, the third is insoluble.

When the newly-expressed juices of vegetables are allowed to stand, a separation takes place in a few minutes. A gelatinous precipitate, commonly of a green tinge, is deposited, and this, when acted on by liquids which remove the colouring matter, leaves a grayish white substance, well known to druggists as the deposite from vegetable juices. This is one of the nitrogenised compounds which serves for the nutrition of animals, and has been named vegetable fibrine. The juice of grapes is especially rich in this constituent, but it is most abundant in the seeds of wheat, and of the cerealia generally. It may be obtained from wheat flour by a mechanical operation, and in a state of tolerable purity; it is then called gluten, but the glutinous property belongs, not to vegetable fibrine, but to a foreign substance, present in small quantity, which is not found in the other cerealia.

The method by which it is obtained sufficiently proves that it is insoluble in water; although we cannot doubt that it was originally dissolved in the vegetable juice, from which it afterwards separated, exactly as fibrine does from blood.

The second nitrogenised compound remains dissolved in the juice after the separation of the fibrine. It does not separate from the juice at the ordinary temperature, but is instantly coagulated when the liquid containing it is heated to the boiling point.

When the clarified juice of nutritious vegetables, such as cauliflower, asparagus, mangel-wurzel, or turnips, is made to boil, a coagulum is formed, which it is absolutely impossible to distinguish from the substance which separates as a coagulum, when the serum of blood, or the white of an egg, diluted with water, are heated to the boiling point. This is vegetable albumen. It is found in the greatest abundance in certain seeds, in nuts, almonds, and others, in which the starch of the gramineae is replaced by oil.

The third nitrogenised constituent of the vegetable food of animals is vegetable caseine. It is chiefly found in the seeds of peas, beans, lentils, and similar leguminous seeds. Like vegetable albumen, it is soluble in water, but differs from it in this, that its solution is not coagulated by heat. When the solution is heated or evaporated, a skin forms on its surface, and the addition of an acid causes a coagulum, just as in animal milk.

These three nitrogenised compounds, vegetable fibrine, albumen, and caseine, are the true nitrogenised constituents of the food of graminivorous animals; all other nitrogenised compounds occurring in plants, are either rejected by animals, as in the case of the characteristic principles of poisonous and medicinal plants, or else they occur in the food in such very small proportion, that they cannot possibly contribute to the increase of mass in the animal body.

The chemical analysis of these three substances has led to the very interesting result that they contain the same organic elements, united in the same proportion by weight; and, what is still more remarkable, that they are identical in composition with the chief constituents of blood, animal fibrine, and albumen. They all three dissolve in concentrated muriatic acid with the same deep purple colour, and even in their physical characters, animal fibrine and albumen are in no respect different from vegetable fibrine and albumen. It is especially to be noticed, that by the phrase, identity of composition, we do not here intend mere similarity, but that even in regard to the presence and relative amount of sulphur, phosphorus, and phosphate of lime, no difference can be observed.

How beautifully and admirably simple, with the aid of these discoveries, appears the process of nutrition in animals, the formation of their organs, in which vitality chiefly resides! Those vegetable principles, which in animals are used to form blood, contain the chief constituents of blood, fibrine and albumen, ready formed, as far as regards their composition. All plants, besides, contain a certain quantity of iron, which reappears in the colouring matter of the blood. Vegetable fibrine and animal fibrine, vegetable albumen and animal albumen, hardly differ, even in form; if these principles be wanting in the food, the nutrition of the animal is arrested; and when they are present, the graminivorous animal obtains in its food the very same principles on the presence of which the nutrition of the carnivora entirely depends.

Vegetables produce in their organism the blood of all animals, for the carnivora, in consuming the blood and flesh of the graminivora, consume, strictly speaking, only the vegetable principles which have served for the nutrition of the latter. Vegetable fibrine and albumen take the form in the stomach of the graminivorous animal as animal fibrine and albumen do in that of the carnivorous animal.

From what has been said, it follows that the development of the animal organism and its growth are dependent on the reception of certain principles identical with the chief constituents of blood.

In this sense we may say that the animal organism gives to the blood only its form; that it is incapable of creating blood out of other substances which do not already contain the chief constituents of that fluid. We cannot, indeed, maintain that the animal organism has no power to form other compounds, for we know that it is capable of producing an extensive series of compounds, differing in composition from the chief constituents of blood; but these last, which form the starting-point of the series, it cannot produce.

The animal organism is a higher kind of vegetable, the development of which begins with those substances with the production of which the life of an ordinary vegetable ends. As soon as the latter has borne seed, it dies, or a period of its life comes to a termination.

In that endless series of compounds, which begins with carbonic acid, ammonia, and water, the sources of the nutrition of vegetables, and includes the most complex constituents of the animal brain, there is no blank, no interruption. The first substance capable of affording nutriment to animals is the last product of the creative energy of vegetables.

The substance of cellular tissue and of membranes, of the brain and nerves, these the vegetable cannot produce.

The seemingly miraculous in the productive agency of vegetables disappears in a great degree, when we reflect that the production of the constituents of blood cannot appear more surprising than the occurrence of the fat of beef and mutton in cocoa beans, of human fat in olive-oil, of the principal ingredient of butter in palm-oil, and of horse fat and train-oil in certain oily seeds.

LETTER IX

My dear Sir,

The facts detailed in my last letter will satisfy you as to the manner in which the increase of mass in an animal, that is, its growth, is accomplished; we have still to consider a most important question, namely, the function performed in the animal system by substances destitute of nitrogen; such as sugar, starch, gum, pectine,

The most extensive class of animals, the graminivora, cannot live without these substances; their food must contain a certain amount of one or more of them, and if these compounds are not supplied, death quickly ensues.

This important inquiry extends also to the constituents of the food of carnivorous animals in the earliest periods of life; for this food also contains substances, which are not necessary for their support in the adult state. The nutrition of the young of carnivora is obviously accomplished by means similar to those by which the graminivora are nourished; their development is dependent on the supply of a fluid, which the body of the mother secretes in the shape of milk.

Milk contains only one nitrogenised constituent, known under the name of caseine; besides this, its chief ingredients are butter (fat), and sugar of milk. The blood of the young animal, its muscular fibre, cellular tissue, nervous matter, and bones, must have derived their origin from the nitrogenised constituent of milk – the caseine; for butter and sugar of milk contain no nitrogen.

Now, the analysis of caseine has led to the result, which, after the details I have given, can hardly excite your surprise, that this substance also is identical in composition with the chief constituents of blood, fibrine and albumen. Nay more – a comparison of its properties with those of vegetable caseine has shown – that these two substances are identical in all their properties; insomuch, that certain plants, such as peas, beans, and lentils, are capable of producing the same substance which is formed from the blood of the mother, and employed in yielding the blood of the young animal.

The young animal, therefore, receives in the form of caseine, – which is distinguished from fibrine and albumen by its great solubility, and by not coagulating when heated, – the chief constituent of the mother's blood. To convert caseine into blood no foreign substance is required, and in the conversion of the mother's blood into caseine, no elements of the constituents of the blood have been separated. When chemically examined, caseine is found to contain a much larger proportion of the earth of bones than blood does, and that in a very soluble form, capable of reaching every part of the body. Thus, even in the earliest period of its life, the development of the organs, in which vitality resides, is, in the carnivorous animal, dependent on the supply of a substance, identical in organic composition with the chief constituents of its blood.

What, then, is the use of the butter and the sugar of milk? How does it happen that these substances are indispensable to life?

Butter and sugar of milk contain no fixed bases, no soda nor potash. Sugar of milk has a composition closely allied to that of the other kinds of sugar, of starch, and of gum; all of them contain carbon and the elements of water, the latter precisely in the proportion to form water.

There is added, therefore, by means of these compounds, to the nitrogenised constituents of food, a certain amount of carbon; or, as in the case of butter, of carbon and hydrogen; that is, an excess of elements, which cannot possibly be employed in the production of blood, because the nitrogenised substances contained in the food already contain exactly the amount of carbon which is required for the production of fibrine and albumen.

In an adult carnivorous animal, which neither gains nor loses weight, perceptibly, from day to day, its nourishment, the waste of organised tissue, and its consumption of oxygen, stand to each other in a well-defined and fixed relation.

The carbon of the carbonic acid given off, with that of the urine; the nitrogen of the urine, and the hydrogen given off as ammonia and water; these elements, taken together, must be exactly equal in weight to the carbon, nitrogen, and hydrogen of the metamorphosed tissues, and since these last are exactly replaced by the food, to the carbon, nitrogen, and hydrogen of the food. Were this not the case, the weight of the animal could not possibly remain unchanged.

But, in the young of the carnivora, the weight does not remain unchanged; on the contrary, it increases from day to day by an appreciable quantity.

This fact presupposes, that the assimilative process in the young animal is more energetic, more intense, than the process of transformation in the existing tissues. If both processes were equally active, the weight of the body could not increase; and were the waste by transformation greater, the weight of the body would decrease.

Now, the circulation in the young animal is not weaker, but, on the contrary, more rapid; the respirations are more frequent; and, for equal bulks, the consumption of oxygen must be greater rather than smaller in the young than in the adult animal. But, since the metamorphosis of organised parts goes on more slowly, there would ensue a deficiency of those substances, the carbon and hydrogen of which are adapted for combination with oxygen; because, in the carnivora, nature has destined the new compounds, produced by the metamorphosis of organised parts, to furnish the necessary resistance to the action of the oxygen, and to produce animal heat. What is wanting for these purposes an Infinite Wisdom has supplied to the young in its natural food.

The carbon and hydrogen of butter, and the carbon of the sugar of milk, no part of either of which can yield blood, fibrine, or albumen, are destined for the support of the respiratory process, at an age when a greater resistance is opposed to the metamorphosis of existing organisms; or, in other words, to the production of compounds, which, in the adult state, are produced in quantity amply sufficient for the purpose of respiration.

The young animal receives the constituents of its blood in the caseine of the milk. A metamorphosis of existing organs goes on, for bile and urine are secreted; the materials of the metamorphosed parts are given off in the form of urine, of carbonic acid, and of water; but the butter and sugar of milk also disappear; they cannot be detected in the faeces.

The butter and sugar of milk are given out in the form of carbonic acid and water, and their conversion into oxidised products furnishes the clearest proof that far more oxygen is absorbed than is required to convert the carbon and hydrogen of the metamorphosed tissues into carbonic acid and water.

The change and metamorphosis of organised tissues going on in the vital process in the young animal, consequently yield, in a given time, much less carbon and hydrogen in the form adapted for the respiratory process than correspond to the oxygen taken up in the lungs. The substance of its organised parts would undergo a more rapid consumption, and would necessarily yield to the action of the oxygen, were not the deficiency of carbon and hydrogen supplied from another source.

The continued increase of mass, or growth, and the free and unimpeded development of the organs in the young animal, are dependent on the presence of foreign substances, which, in the nutritive process, have no other function than to protect the newly-formed organs from the action of the oxygen. The elements of these substances unite with the oxygen; the organs themselves could not do so without being consumed; that is, growth, or increase of mass in the body, – the consumption of oxygen remaining the same, – would be utterly impossible.

The preceding considerations leave no doubt as to the purpose for which Nature has added to the food of the young of carnivorous mammalia substances devoid of nitrogen, which their organism cannot employ for nutrition, strictly so called, that is, for the production of blood; substances which may be entirely dispensed with in their nourishment in the adult state. In the young of carnivorous birds, the want of all motion is an obvious cause of diminished waste in the organised parts; hence, milk is not provided for them.

The nutritive process in the carnivora thus presents itself under two distinct forms; one of which we again meet with in the graminivora.

In graminivorous animals. we observe, that during their whole life, their existence depends on a supply of substances having a composition identical with that of sugar of milk, or closely resembling it. Everything that they consume as food contains a certain quantity of starch, gum, or sugar, mixed with other matters.

The function performed in the vital process of the graminivora by these substances is indicated in a very clear and convincing manner, when we take into consideration the very small relative amount of the carbon which these animals consume in the nitrogenised constituents of their food, which bears no proportion whatever to the oxygen absorbed through the skin and lungs.

A horse, for example, can be kept in perfectly good condition, if he obtain as food 15 lbs. of hay and 4 1/2 lbs. of oats daily. If we now calculate the whole amount of nitrogen in these matters, as ascertained by analysis (1 1/2 per cent. in the hay, 2.2 per cent. in the oats), in the form of blood, that is, as fibrine and albumen, with the due proportion of water in blood (80 per cent.), the horse receives daily no more than 4 1/2 oz. of nitrogen, corresponding to about 8 lbs. of blood. But along with this nitrogen, that is, combined with it in the form of fibrine or albumen, the animal receives only about 14 1/2 oz. of carbon.

Without going further into the calculation, it will readily be admitted, that the volume of air inspired and expired by a horse, the quantity of oxygen consumed, and, as a necessary consequence, the amount of carbonic acid given out by the animal, are much greater than in the respiratory process in man. But an adult man consumes daily abut 14 oz. of carbon, and the determination of Boussingault, according to which a horse expires 79 oz. daily, cannot be very far from the truth.

In the nitrogenised constituents of his food, therefore, the horse receives rather less than the fifth part of the carbon which his organism requires for the support of the respiratory process; and we see that the wisdom of the Creator has added to his food the four-fifths which are wanting, in various forms, as starch, sugar, with which the animal must be supplied, or his organism will be destroyed by the action of the oxygen.

It is obvious, that in the system of the graminivora, whose food contains so small a portion, relatively, of the constituents of the blood, the process of metamorphosis in existing tissues, and consequently their restoration or reproduction, must go on far less rapidly than in the carnivora. Were this not the case, a vegetation a thousand times more luxuriant than the actual one would not suffice for their nourishment. Sugar, gum, and starch, would no longer be necessary to support life in these animals, because, in that case, the products of the waste, or metamorphosis of the organised tissues, would contain enough carbon to support the respiratory process.

LETTER X

My dear Sir,

Let me now apply the principles announced in the preceding letters to the circumstances of our own species. Man, when confined to animal food, requires for his support and nourishment extensive sources of food, even more widely extended than the lion and tiger, because, when he has the opportunity, he kills without eating.

A nation of hunters, on a limited space, is utterly incapable of increasing its numbers beyond a certain point, which is soon attained. The carbon necessary for respiration must be obtained from the animals, of which only a limited number can live on the space supposed. These animals collect from plants the constituents of their organs and of their blood, and yield them, in turn, to the savages who live by the chase alone. They, again, receive this food unaccompanied by those compounds, destitute of nitrogen, which, during the life of the animals, served to support the respiratory process. In such men, confined to an animal diet, it is the carbon of the flesh and of the blood which must take the place of starch and sugar.

But 15 lbs. of flesh contain no more carbon than 4 lbs. of starch, and while the savage with one animal and an equal weight of starch should maintain life and health for a certain number of days, he would be compelled, if confined to flesh alone, in order to procure the carbon necessary for respiration, during the same time, to consume five such animals.

It is easy to see, from these considerations, how close the connection is between agriculture and the multiplication of the human species. The cultivation of our crops has ultimately no other object than the production of a maximum of those substances which are adapted for assimilation and respiration, in the smallest possible space. Grain and other nutritious vegetables yield us, not only in starch, sugar, and gum, the carbon which protects our organs from the action of oxygen, and produces in the organism the heat which is essential to life, but also in the form of vegetable fibrine, albumen, and caseine, our blood, from which the other parts of our body are developed.

Man, when confined to animal food, respires, like the carnivora, at the expense of the matters produced by the metamorphosis of organised tissues; and, just as the lion, tiger, hyaena, in the cages of a menagerie, are compelled to accelerate the waste of the organised tissues by incessant motion, in order to furnish the matter necessary for respiration, so, the savage, for the very same object, is forced to make the most laborious exertions, and go through a vast amount of muscular exercise. He is compelled to consume force merely in order to supply matter for respiration.

Cultivation is the economy of force. Science teaches us the simplest means of obtaining the greatest effect with the smallest expenditure of power, and with given means to produce a maximum of force. The unprofitable exertion of power, the waste of force in agriculture, in other branches of industry, in science, or in social economy, is characteristic of the savage state, or of the want of knowledge.

In accordance with what I have already stated, you will perceive that the substances of which the food of man is composed may be divided into two classes; into nitrogenised and non-nitrogenised. The former are capable of conversion into blood; the latter are incapable of this transformation.

Out of those substances which are adapted to the formation of blood, are formed all the organised tissues. The other class of substances, in the normal state of health, serve to support the process of respiration. The former may be called the plastic elements of nutrition; the latter, elements of respiration.

Among the former we reckon—

Vegetable fibrine.

Vegetable albumen.

Vegetable caseine.

Animal flesh.

Animal blood.

Among the elements of respiration in our food, are—

Fat. Pectine.

Starch. Bassorine.

Gum. Wine.

Cane sugar. Beer.

Grape sugar. Spirits.

Sugar of milk.

The most recent and exact researches have established as a universal fact, to which nothing yet known is opposed, that the nitrogenised constituents of vegetable food have a composition identical with that of the constituents of the blood.

No nitrogenised compound, the composition of which differs from that of fibrine, albumen, and caseine, is capable of supporting the vital process in animals.

The animal organism unquestionably possesses the power of forming, from the constituents of its blood, the substance of its membranes and cellular tissue, of the nerves and brain, and of the organic part of cartilages and bones. But the blood must be supplied to it perfect in everything but its form – that is, in its chemical composition. If this be not done, a period is rapidly put to the formation of blood, and consequently to life.

This consideration enables us easily to explain how it happens that the tissues yielding gelatine or chondrine, as, for example, the gelatine of skin or of bones, are not adapted for the support of the vital process; for their composition is different from that of fibrine or albumen. It is obvious that this means nothing more than that those parts of the animal organism which form the blood do not possess the power of effecting a transformation in the arrangement of the elements of gelatine, or of those tissues which contain it. The gelatinous tissues, the gelatine of the bones, the membranes, the cells and the skin suffer, in the animal body, under the influence of oxygen and moisture, a progressive alteration; a part of these tissues is separated, and must be restored from the blood; but this alteration and restoration are obviously confined within very narrow limits.

While, in the body of a starving or sick individual, the fat disappears and the muscular tissue takes once more the form of blood, we find that the tendons and membranes retain their natural condition, and the limbs of the dead body their connections, which depend on the gelatinous tissues.

On the other hand, we see that the gelatine of bones devoured by a dog entirely disappears, while only the bone earth is found in his excrements. The same is true of man, when fed on food rich in gelatine, as, for example, strong soup. The gelatine is not to be found either in the urine or in the faeces, and consequently must have undergone a change, and must have served some purpose in the animal economy. It is clear that the gelatine must be expelled from the body in a form different from that in which it was introduced as food.

When we consider the transformation of the albumen of the blood into a part of an organ composed of fibrine, the identity in composition of the two substances renders the change easily conceivable. Indeed we find the change of a dissolved substance into an insoluble organ of vitality, chemically speaking, natural and easily explained, on account of this very identity of composition. Hence the opinion is not unworthy of a closer investigation, that gelatine, when taken in the dissolved state, is again converted, in the body, into cellular tissue, membrane and cartilage; that it may serve for the reproduction of such parts of these tissues as have been wasted, and for their growth.

And when the powers of nutrition in the whole body are affected by a change of the health, then, even should the power of forming blood remain the same, the organic force by which the constituents of the blood are transformed into cellular tissue and membranes must necessarily be enfeebled by sickness. In the sick man, the intensity of the vital force, its power to produce metamorphoses, must be diminished as well in the stomach as in all other parts of the body. In this condition, the uniform experience of practical physicians shows that gelatinous matters in a dissolved state exercise a most decided influence on the state of the health. Given in a form adapted for assimilation, they serve to husband the vital force, just as may be done, in the case of the stomach, by due preparation of the food in general.

Brittleness in the bones of graminivorous animals is clearly owing to a weakness in those parts of the organism whose function it is to convert the constituents of the blood into cellular tissue and membrane; and if we can trust to the reports of physicians who have resided in the East, the Turkish women, in their diet of rice, and in the frequent use of enemata of strong soup, have united the conditions necessary for the formation both of cellular tissue and of fat.

LETTER XI

My dear Sir,

In the immense, yet limited expanse of the ocean, the animal and vegetable kingdoms are mutually dependent upon, and successive to each other. The animals obtain their constituent elements from the plants, and restore them to the water in their original form, when they again serve as nourishment to a new generation of plants.

The oxygen which marine animals withdraw in their respiration from the air, dissolved in sea water, is returned to the water by the vital processes of sea plants; that air is richer in oxygen than atmospheric air, containing 32 to 33 per cent. Oxygen, also, combines with the products of the putrefaction of dead animal bodies, changes their carbon into carbonic acid, their hydrogen into water, and their nitrogen assumes again the form of ammonia.

Thus we observe in the ocean a circulation takes place without the addition or subtraction of any element, unlimited in duration, although limited in extent, inasmuch as in a confined space the nourishment of plants exists in a limited quantity.

We well know that marine plants cannot derive a supply of humus for their nourishment through their roots. Look at the great sea-tang, the Fucus giganteus: this plant, according to Cook, reaches a height of 360 feet, and a single specimen, with its immense ramifications, nourishes thousands of marine animals, yet its root is a small body, no larger than the fist. What nourishment can this draw from a naked rock, upon the surface of which there is no perceptible change? It is quite obvious that these plants require only a hold, – a fastening to prevent a change of place, – as a counterpoise to their specific gravity, which is less than that of the medium in which they float. That medium provides the necessary nourishment, and presents it to the surface of every part of the plant. seawater contains not only carbonic acid and ammonia, but the alkaline and earthy phosphates and carbonates required by these plants for their growth, and which we always find as constant constituents of their ashes.

All experience demonstrates that the conditions of the existence of marine plants are the same which are essential to terrestrial plants. But the latter do not live like sea-plants, in a medium which contains all their elements and surrounds with appropriate nourishment every part of their organs; on the contrary, they require two media, of which one, namely the soil, contains those essential elements which are absent from the medium surrounding them, i.e. the atmosphere.

Is it possible that we could ever be in doubt respecting the office which the soil and its component parts subserve in the existence and growth of vegetables? – that there should have been a time when the mineral elements of plants were not regarded as absolutely essential to their vitality? Has not the same circulation been observed on the surface of the earth which we have just contemplated in the ocean, – the same incessant change, disturbance and restitution of equilibrium?

Experience in agriculture shows that the production of vegetables on a given surface increases with the supply of certain matters, originally parts of the soil which had been taken up from it by plants – the excrements of man and animals. These are nothing more than matters derived from vegetable food, which in the vital processes of animals, or after their death, assume again the form under which they originally existed, as parts of the soil. Now, we know that the atmosphere contains none of these substances, and therefore can replace none; and we know that their removal from a soil destroys its fertility, which may be restored and increased by a new supply.

Is it possible, after so many decisive investigations into the origin of the elements of animals and vegetables, the use of the alkalies, of lime and the phosphates, any doubt can exist as to the principles upon which a rational agriculture depends? Can the art of agriculture be based upon anything but the restitution of a disturbed equilibrium? Can it be imagined that any country, however rich and fertile, with a flourishing commerce, which for centuries exports its produce in the shape of grain and cattle, will maintain its fertility, if the same commerce does not restore, in some form of manure, those elements which have been removed from the soil, and which cannot be replaced by the atmosphere? Must not the same fate await every such country which has actually befallen the once prolific soil of Virginia, now in many parts no longer able to grow its former staple productions – wheat and tobacco?

In the large towns of England the produce both of English and foreign agriculture is largely consumed; elements of the soil indispensable to plants do not return to the fields, – contrivances resulting from the manners and customs of English people, and peculiar to them, render it difficult, perhaps impossible, to collect the enormous quantity of the phosphates which are daily, as solid and liquid excrements, carried into the rivers. These phosphates, although present in the soil in the smallest quantity, are its most important mineral constituents. It was observed that many English fields exhausted in that manner immediately doubled their produce, as if by a miracle, when dressed with bone earth imported from the Continent. But if the export of bones from Germany is continued to the extent it has hitherto reached, our soil must be gradually exhausted, and the extent of our loss may be estimated, by considering that one pound of bones contains as much phosphoric acid as a hundredweight of grain.

The imperfect knowledge of Nature and the properties and relations of matter possessed by the alchemists gave rise, in their time, to an opinion that metals as well as plants could be produced from a seed. The regular forms and ramifications seen in crystals, they imagined to be the leaves and branches of metal plants; and as they saw the seed of plants grow, producing root, stem and leaves, and again blossoms, fruit and seeds, apparently without receiving any supply of appropriate material, they deemed it worthy of zealous inquiry to discover the seed of gold, and the earth necessary for its development. If the metal seeds were once obtained, might they not entertain hopes of their growth?

Such ideas could only be entertained when nothing was known of the atmosphere, and its participation with the earth, in administering to the vital processes of plants and animals. Modern chemistry indeed produces the elements of water, and, combining them, forms water anew; but it does not create those elements – it derives them from water; the new-formed artificial water has been water before.

Many of our farmers are like the alchemists of old, – they are searching for the miraculous seed, – the means, which, without any further supply of nourishment to a soil scarcely rich enough to be sprinkled with indigenous plants, shall produce crops of grain a hundred-fold.

The experience of centuries, nay, of thousands of years, is insufficient to guard men against these fallacies; our only security from these and similar absurdities must be derived from a correct knowledge of scientific principles.

In the first period of natural philosophy, organic life was supposed to be derived from water only; afterwards, it was admitted that certain elements derived from the air must be superadded to the water; but we now know that other elements must be supplied by the earth, if plants are to thrive and multiply.

The amount of materials contained in the atmosphere, suited to the nourishment of plants, is limited; but it must be abundantly sufficient to cover the whole surface of the earth with a rich vegetation. Under the tropics, and in those parts of our globe where the most genial conditions of fertility exist, – a suitable soil, a moist atmosphere, and a high temperature, – vegetation is scarcely limited by space; and, where the soil is wanting, it is gradually supplied by the decaying leaves, bark and branches of plants. It is obvious there is no deficiency of atmospheric nourishment for plants in those regions, nor are these wanting in our own cultivated fields: all that plants require for their development is conveyed to them by the incessant motions of the atmosphere. The air between the tropics contains no more than that of the arctic zones; and yet how different is the amount of produce of an equal surface of land in the two situations!

This is easily explicable. All the plants of tropical climates, the oil and wax palms, the sugar cane, contain only a small quantity of the elements of the blood necessary to the nutrition of animals, as compared with our cultivated plants. The tubers of the potato in Chili, its native country, where the plant resembles a shrub, if collected from an acre of land, would scarcely suffice to maintain an Irish family for a single day (Darwin). The result of cultivation in those plants which serve as food, is to produce in them those constituents of the blood. In the absence of the elements essential to these in the soil, starch, sugar and woody fibre, are perhaps formed; but no vegetable fibrine, albumen, or caseine. If we intend to produce on a given surface of soil more of these latter matters than the plants can obtain from the atmosphere or receive from the soil of the same surface in its uncultivated and normal state, we must create an artificial atmosphere, and add the needed elements to the soil.

The nourishment which must be supplied in a given time to different plants, in order to admit a free and unimpeded growth, is very unequal.

On pure sand, on calcareous soil, on naked rocks, only a few genera of plants prosper, and these are, for the most part, perennial plants. They require, for their slow growth, only such minute quantities of mineral substances as the soil can furnish, which may be totally barren for other species. Annual, and especially summer plants, grow and attain their perfection in a comparatively short time; they therefore do not prosper on a soil which is poor in those mineral substances necessary to their development. To attain a maximum in height in the short period of their existence, the nourishment contained in the atmosphere is not sufficient. If the end of cultivation is to be obtained, we must create in the soil an artificial atmosphere of carbonic acid and ammonia; and this surplus of nourishment, which the leaves cannot appropriate from the air, must be taken up by the corresponding organs, i.e. the roots, from the soil. But the ammonia, together with the carbonic acid, are alone insufficient to become part of a plant destined to the nourishment of animals. In the absence of the alkalies, the phosphates and other earthy salts, no vegetable fibrine, no vegetable caseine, can be formed. The phosphoric acid of the phosphate of lime, indispensable to the cerealia and other vegetables in the formation of their seeds, is separated as an excrement, in great quantities, by the rind and barks of ligneous plants.

How different are the evergreen plants, the cacti, the mosses, the ferns, and the pines, from our annual grasses, the cerealia and leguminous vegetables! The former, at every time of the day during winter and summer, obtain carbon through their leaves by absorbing carbonic acid which is not furnished by the barren soil on which they grow; water is also absorbed and retained by their coriaceous or fleshy leaves with great force. They lose very little by evaporation, compared with other plants. On the other hand, how very small is the quantity of mineral substances which they withdraw from the soil during their almost constant growth in one year, in comparison with the quantity which one crop of wheat of an equal weight receives in three months!

It is by means of moisture that plants receive the necessary alkalies and salts from the soil. In dry summers a phenomenon is observed, which, when the importance of mineral elements to the life of a plant was unknown, could not be explained. The leaves of plants first developed and perfected, and therefore nearer the surface of the soil, shrivel up and become yellow, lose their vitality, and fall off while the plant is in an active state of growth, without any visible cause. This phenomenon is not seen in moist years, nor in evergreen plants, and but rarely in plants which have long and deep roots, nor is it seen in perennials in autumn and winter.

The cause of this premature decay is now obvious. The perfectly-developed leaves absorb continually carbonic acid and ammonia from the atmosphere, which are converted into elements of new leaves, buds, and shoots; but this metamorphosis cannot be effected without the aid of the alkalies, and other mineral substances. If the soil is moist, the latter are continually supplied to an adequate amount, and the plant retains its lively green colour; but if this supply ceases from a want of moisture to dissolve the mineral elements, a separation takes place in the plant itself. The mineral constituents of the juice are withdrawn from the leaves already formed, and are used for the formation of the young shoots; and as soon as the seeds are developed, the vitality of the leaves completely ceases. These withered leaves contain only minute traces of soluble salts, while the buds and shoots are very rich in them.

On the other hand, it has been observed, that where a soil is too highly impregnated with soluble saline materials, these are separated upon the surface of the leaves. This happens to culinary vegetables especially, whose leaves become covered with a white crust. In consequence of these exudations the plant sickens, its organic activity decreases, its growth is disturbed; and if this state continues long, the plant dies. This is most frequently seen in foliaceous plants, the large surfaces of which evaporate considerable quantities of water. Carrots, pumpkins, peas, are frequently thus diseased, when, after dry weather, the plant being near its full growth, the soil is moistened by short showers, followed again by dry weather. The rapid evaporation carries off the water absorbed by the root, and this leaves the salts in the plant in a far greater quantity than it can assimilate. These salts effloresce upon the surface of the leaves, and if they are herbaceous and juicy, produce an effect upon them as if they had been watered with a solution containing a greater quantity of salts than their organism can bear.

Of two plants of the same species, this disease befalls that which is nearest its perfection; if one should have been planted later, or be more backward in its development, the same external cause which destroys the one will contribute to the growth of the other.

LETTER XII

My dear Sir,

Having now occupied several letters with the attempt to unravel, by means of chemistry, some of the most curious functions of the animal body, and, as I hope, made clear to you the distinctions between the two kinds of constituent elements in food, and the purposes they severally subserve in sustaining life, let me now direct your attention to a scarcely less interesting and equally important subject – the means of obtaining from a given surface of the earth the largest amount of produce adapted to the food of man and animals.

Agriculture is both a science and an art. The knowledge of all the conditions of the life of vegetables, the origin of their elements, and the sources of their nourishment, forms its scientific basis.

From this knowledge we derive certain rules for the exercise of the ART, the principles upon which the mechanical operations of farming depend, the usefulness or necessity of these for preparing the soil to support the growth of plants, and for removing every obnoxious influence. No experience, drawn from the exercise of the art, can be opposed to true scientific principles, because the latter should include all the results of practical operations, and are in some instances solely derived therefrom. Theory must correspond with experience, because it is nothing more than the reduction of a series of phenomena to their last causes.

A field in which we cultivate the same plant for several successive years becomes barren for that plant in a period varying with the nature of the soil: in one field it will be in three, in another in seven, in a third in twenty, in a fourth in a hundred years. One field bears wheat, and no peas; another beans or turnips, but no tobacco; a third gives a plentiful crop of turnips, but will not bear clover. What is the reason that a field loses its fertility for one plant, the same which at first flourished there? What is the reason one kind of plant succeeds in a field where another fails?

These questions belong to Science.

What means are necessary to preserve to a field its fertility for one and the same plant? – what to render one field fertile for two, for three, for all plants?

These last questions are put by Art, but they cannot be answered by Art.

If a farmer, without the guidance of just scientific principles, is trying experiments to render a field fertile for a plant which it otherwise will not bear, his prospect of success is very small. Thousands of farmers try such experiments in various directions, the result of which is a mass of practical experience forming a method of cultivation which accomplishes the desired end for certain places; but the same method frequently does not succeed, it indeed ceases to be applicable to a second or third place in the immediate neighbourhood. How large a capital, and how much power, are wasted in these experiments! Very different, and far more secure, is the path indicated by SCIENCE; it exposes us to no danger of failing, but, on the contrary, it furnishes us with every guarantee of success. If the cause of failure – of barrenness in the soil for one or two plants – has been discovered, means to remedy it may readily be found.

The most exact observations prove that the method of cultivation must vary with the geognostical condition of the subsoil. In basalt, graywacke, porphyry, sandstone, limestone, are certain elements indispensable to the growth of plants, and the presence of which renders them fertile. This fully explains the difference in the necessary methods of culture for different places; since it is obvious that the essential elements of the soil must vary with the varieties of composition of the rocks, from the disintegration of which they originated.

Wheat, clover, turnips, for example, each require certain elements from the soil; they will not flourish where the appropriate elements are absent. Science teaches us what elements are essential to every species of plants by an analysis of their ashes. If therefore a soil is found wanting in any of those elements, we discover at once the cause of its barrenness, and its removal may now be readily accomplished.

The empiric attributes all his success to the mechanical operations of agriculture; he experiences and recognises their value, without inquiring what are the causes of their utility, their mode of action: and yet this scientific knowledge is of the highest importance for regulating the application of power and the expenditure of capital, – for insuring its economical expenditure and the prevention of waste. Can it be imagined that the mere passing of the ploughshare or the harrow through the soil – the mere contact of the iron – can impart fertility miraculously? Nobody, perhaps, seriously entertains such an opinion. Nevertheless, the modus operandi of these mechanical operations is by no means generally understood. The fact is quite certain, that careful ploughing exerts the most favourable influence: the surface is thus mechanically divided, changed, increased, and renovated; but the ploughing is only auxiliary to the end sought.

In the effects of time, in what in Agriculture are technically called fallows – the repose of the fields – we recognise by science certain chemical actions, which are continually exercised by the elements of the atmosphere upon the whole surface of our globe. By the action of its oxygen and its carbonic acid, aided by water, rain, changes of temperature, certain elementary constituents of rocks, or of their ruins, which form the soil capable of cultivation, are rendered soluble in water, and conseqently become separable from all their insoluble parts.

These chemical actions, poetically denominates the "tooth of time," destroy all the works of man, and gradually reduce the hardest rocks to the condition of dust. By their influence the necessary elements of the soil become fitted for assimilation by plants; and it is precisely the end which is obtained by the mechanical operations of farming. They accelerate the decomposition of the soil, in order to provide a new generation of plants with the necesary elements in a condition favourable to their assimilation. It is obvious that the rapidity of the decomposition of a solid body must increase with the extension of its surface; the more points of contact we offer in a given time to the external chemical agent, the more rapid will be its action.

The chemist, in order to prepare a mineral for analysis, to decompose it, or to increase the solubility of its elements, proceeds in the same way as the farmer deals with his fields – he spares no labour in order to reduce it to the finest powder; he separates the impalpable from the coarser parts by washing, and repeats his mechanical bruising and trituration, being assured his whole process will fail if he is inattentive to this essential and preliminary part of it.

The influence which the increase of surface exercises upon the disintegration of rocks, and upon the chemical action of air and moisture, is strikingly illustrated upon a large scale in the operations pursued in the goldmines of Yaquil, in Chili. These are described in a very interesting manner by Darwin. The rock containing the gold ore is pounded by mills into the finest powder; this is subjected to washing, which separates the lighter particles from the metallic; the gold sinks to the bottom, while a stream of water carries away the lighter earthy parts into ponds, where it subsides to the bottom as mud. When this deposit has gradually filled up the pond, this mud is taken out and piled in heaps, and left exposed to the action of the atmosphere and moisture. The washing completely removes all the soluble part of the disintegrated rock; the insoluble part, moreover, cannot undergo any further change while it is covered with water, and so excluded from the influence of the atmosphere at the bottom of the pond. But being exposed at once to the air and moisture, a powerful chemical action takes place in the whole mass, which becomes indicated by an efflorescence of salts covering the whole surface of the heaps in considerable quantity. After being exposed for two or three years, the mud is again subjected to the same process of washing, and a considerable quantity of gold is obtained, this having been separated by the chemical process of decomposition in the mass. The exposure and washing of the same mud is repeated six or seven times, and at every washing it furnishes a new quantity of gold, although its amount diminishes every time.

Precisely similar is the chemical action which takes place in the soil of our fields; and we accelerate and increase it by the mechanical operations of our agriculture. By these we sever and extend the surface, and endeavour to make every atom of the soil accessible to the action of the carbonic acid and oxygen of the atmosphere. We thus produce a stock of soluble mineral substances, which serves as nourishment to a new generation of plants, materials which are indispensable to their growth and prosperity.

LETTER XIII

My dear Sir,

Having in my last letter spoken of the general principles upon which the science and art of agriculture must be based, let me now direct your attention to some of those particulars between chemistry and agriculture, and demonstrate the impossibility of perfecting the important art of rearing food for man and animals, without a profound knowledge of our science.

All plants cultivated as food require for their healthy sustenance the alkalies and alkaline earths, each in a certain proportion; and in addition to these, the cerealia do not succeed in a soil destitute of silica in a soluble condition. The combinations of this substance found as natural productions, namely, the silicates, differ greatly in the degree of facility with which they undergo decomposition, in consequence of the unequal resistance opposed by their integral parts to the dissolving power of the atmospheric agencies. Thus the granite of Corsica degenerates into a powder in a time which scarcely suffices to deprive the polished granite of Heidelberg of its lustre.

Some soils abound in silicates so readily decomposable, that in every one or two years, as much silicate of potash becomes soluble and fitted for assimilation as is required by the leaves and straw of a crop of wheat. In Hungary, extensive districts are not uncommon where wheat and tobacco have been grown alternately upon the same soil for centuries, the land never receiving back any of those mineral elements which were withdrawn in the grain and straw. On the other hand, there are fields in which the necessary amount of soluble silicate of potash for a single crop of wheat is not separated from the insoluble masses in the soil in less than two, three, or even more years.

The term fallow, in Agriculture, designates that period in which the soil, left to the influence of the atmosphere, becomes enriched with those soluble mineral constituents. Fallow, however, does not generally imply an entire cessation of cultivation, but only an interval in the growth of the cerealia. That store of silicates and alkalies which is the principal condition of their success is obtained, if potatoes or turnips are grown upon the same fields in the intermediate periods, since these crops do not abstract a particle of silica, and therefore leave the field equally fertile for the following crop of wheat.

The preceding remarks will render it obvious to you, that the mechanical working of the soil is the simplest and cheapest method of rendering the elements of nutrition contained in it accessible to plants.

But it may be asked, Are there not other means of decomposing the soil besides its mechanical subdivision? – are there not substances, which by their chemical operation will equally well or better render its constituents suitable for entering into vegetable organisms? Yes: we certainly possess such substances, and one of them, namely, quicklime, has been employed for the last century past in England for this purpose; and it would be difficult to find a substance better adapted to this service, as it is simple, and in almost all localities cheap and easily accessible.

In order to obtain correct views respecting the effect of quicklime upon the soil, let me remind you of the first process employed by the chemist when he is desirous of analysing a mineral, and for this purpose wishes to bring its elements into a soluble state. Let the mineral to be examined be, for instance, feldspar; this substance, taken alone, even when reduced to the finest powder, requires for its solution to be treated with an acid for weeks or months; but if we first mix it with quicklime, and expose the mixture to a moderately strong heat, the lime enters into chemical combination with certain elements of the feldspar, and its alkali (potass) is set free. And now the acid, even without heat, dissolves not only the lime, but also so much of the silica of the feldspar as to form a transparent jelly. The same effect which the lime in this process, with the aid of heat, exerts upon the feldspar, it produces when it is mixed with the alkaline argillaceous silicates, and they are for a long time kept together in a moist state.

Common potters' clay, or pipe-clay, diffused through water, and added to milk of lime, thickens immediately upon mixing; and if the mixture is kept for some months, and then treated with acid, the clay becomes gelatinous, which would not occur without the admixture with the lime. The lime, in combining with the elements of the clay, liquifies it; and, what is more remarkable, liberates the greater part of its alkalies. These interesting facts were first observed by Fuchs, at Munich: they have not only led to a more intimate knowledge of the nature and properties of the hydraulic cements, but, what is far more important, they explain the effects of caustic lime upon the soil, and guide the agriculturist in the application of an invaluable means of opening it, and setting free its alkalies – substances so important, nay, so indispensable to his crops.

In the month of October the fields of Yorkshire and Oxfordshire look as it they were covered with snow. Whole square miles are seen whitened over with quicklime, which during the moist winter months, exercises its beneficial influence upon the stiff, clayey soil, of those counties.

According to the humus theory, quicklime ought to exert the most noxious influence upon the soil, because all organic matters contained in it are destroyed by it, and rendered incapable of yielding their humus to a new vegetation. The facts are indeed directly contrary to this now abandoned theory: the fertility of the soil is increased by the lime. The cerealia require the alkalies and alkaline silicates, which the action of the lime renders fit for assimilation by the plants. If, in addition to these, there is any decaying organic matter present in the soil supplying carbonic acid, it may facilitate their development; but it is not essential to their growth. If we furnish the soil with ammonia, and the phosphates, which are indispensable to the cerealia, with the alkaline silicates, we have all the conditions necessary to ensure an abundant harvest. The atmosphere is an inexhaustible store of carbonic acid.

A no less favourable influence than that of lime is exercised upon the soil of peaty land by the mere act of burning it: this greatly enhances its fertility. We have not long been acquainted with the remarkable change which the properties of clay undergo by burning. The observation was first made in the process of analysing the clay silicates. Many of these, in their natural state, are not acted on by acids, but they become perfectly soluble if heated to redness before the application of the acid. This property belongs to potters' clay, pipe-clay, loam, and many different modifications of clay in soils. In their natural state they may be boiled in concentrated sulphuric acid, without sensible change; but if feebly burned, as is done with the pipe-clay in many alum manufactories, they dissolve in the acid with the greatest facility, the contained silica being separated like jelly in a soluble state. Potters' clay belongs to the most sterile kinds of soil, and yet it contains within itself all the constituent elements essential to a most luxurious growth of plants; but their mere presence is insufficient to secure this end. The soil must be accessible to the atmosphere, to its oxygen, to its carbonic acid; these must penetrate it, in order to secure the conditions necessary to a happy and vigorous development of the roots. The elements present must be brought into that peculiar state of combination which will enable them to enter into plants. Plastic clay is wanting in these properties; but they are imparted to it by a feeble calcination.

At Hardwicke Court, near Gloucester, I have seen a garden (Mr. Baker's) consisting of a stiff clay, which was perfectly sterile, become by mere burning extremely fertile. The operation was extended to a depth of three feet. This was an expensive process, certainly; but it was effectual.

The great difference in the properties of burnt and unburnt clay is illustrated by what is seen in brick houses, built in moist situations. In the town of Flanders, for instance, where most buildings are of brick, efflorescences of salts cover the surfaces of the walls, like a white nap, within a few days after they are erected. If this saline incrustation is washed away by the rain, it soon reappears; and this is even observed on walls which, like the gateway of Lisle, have been erected for centuries. These saline incrustations consist of carbonates and sulphates, with alkaline bases;

and it is well known these act an important part in vegetation. The influence of lime in their production is manifested by their appearing first at the place where the mortar and brick come into contact.

It will now be obvious to you, that in a mixture of clay with lime, all the conditions exist for the solution of the silicated clay, and the solubility of the alkaline silicates. The lime gradually dissolving in water charged with carbonic acid, acts like milk of lime upon the clay. This explains also the favourable influence which marl (by which term all those varieties of clay rich in chalk are designated) exerts upon most kinds of soil. There are marly soils which surpass all others in fertility for all kinds of plants; but I believe marl in a burnt state must be far more effective, as well as other materials possessing a similar composition; as, for instance, those species of limestone which are adapted to the preparation of hydraulic cements, – for these carry to the soil not only the alkaline bases useful to plants, but also silica in a state capable of assimilation.

The ashes of coals and lignite are also excellent means of ameliorating the soil, and they are used in many places for this purpose. The most suitable may be readily known by their property of forming a gelatinous mass when treated with acids, or by becoming, when mixed with cream of lime, like hydraulic cement, – solid and hard as stone.

I have now, I trust, explained to your satisfaction, that the mechanical operations of agriculture – the application of lime and chalk to lands, and the burning of clay – depend upon one and the same scientific principle: they are means of accelerating the decomposition of the alkaline clay silicates, in order to provide plants, at the beginning of a new vegetation, with certain inorganic matters indispensable for their nutrition.

LETTER XIV

My dear Sir,

I treated, in my last letter, of the means of improving the condition of the soil for agricultural purposes by mechanical operations and mineral agents. I have now to speak of the uses and effects of animal exuviae, and vegetable matters or manures – properly so called.

In order to understand the nature of these, and the peculiarity of their influence upon our fields, it is highly important to keep in mind the source whence they are derived.

It is generally known, that if we deprive an animal of food, the weight of its body diminishes during every moment of its existence. If this abstinence is continued for some time, the diminution becomes apparent to the eye; all the fat of the body disappears, the muscles decrease in firmness and bulk, and, if the animal is allowed to die starved, scarcely anything but skin, tendon, and bones, remain. This emaciation which occurs in a body otherwise healthy, demonstrates to us, that during the life of an animal every part of its living substance is undergoing a perpetual change; all its component parts, assuming the form of lifeless compounds, are thrown off by the skin, lungs, and urinary system, altered more or less by the secretory organs. This change in the living body is intimately connected with the process of respiration; it is, in truth, occasioned by the oxygen of the atmosphere in breathing, which combines with all the various matters within the body. At every inspiration a quantity of oxygen passes into the blood in the lungs, and unites with its elements; but although the weight of the oxygen thus daily entering into the body amounts to 32 or more ounces, yet the weight of the body is not thereby increased. Exactly as much oxygen as is imbibed in inspiration passes off in expiration, in the form of carbonic acid and water; so that with every breath the amount of carbon and hydrogen in the body is diminished. But the emaciation – the loss of weight by starvation – does not simply depend upon the separation of the carbon and hydrogen; but all the other substances which are in combination with these elements in the living tissues pass off in the secretions. The nitrogen undergoes a change, and is thrown out of the system by the kidneys. Their secretion, the urine, contains not only a compound rich in nitrogen, namely urea, but the sulphur of the tissues in the form of a sulphate, all the soluble salts of the blood and animal fluids, common salt, the phosphates, soda and potash. The carbon and hydrogen of the blood, of the muscular fibre, and of all the animal tissues which can undergo change, return into the atmosphere. The nitrogen, and all the soluble inorganic elements are carried to the earth in the urine.

These changes take place in the healthy animal body during every moment of life; a waste and loss of substance proceeds continually; and if this loss is to be restored, and the original weight and substance repaired, an adequate supply of materials must be furnished, from whence the blood and wasted tissues may be regenerated. This supply is obtained from the food.

In an adult person in a normal or healthy condition, no sensible increase or decrease of weight occurs from day to day. In youth the weight of the body increases, whilst in old age it decreases. There can be no doubt that in the adult, the food has exactly replaced the loss of substance: it has supplied just so much carbon, hydrogen, nitrogen, and other elements, as have passed through the skin, lungs, and urinary organs. In youth the supply is greater than the waste. Part of the elements of the food remain to augment the bulk of the body. In old age the waste is greater than the supply, and the body diminishes. It is unquestionable, that, with the exception of a certain quantity of carbon and hydrogen, which are secreted through the skin and lungs, we obtain, in the solid and fluid excrements of man and animals, all the elements of their food.

We obtain daily, in the form of urea, all the nitrogen taken in the food both of the young and the adult; and further, in the urine, the whole amount of the alkalies, soluble phosphates and sulphates, contained in all the various aliments. In the solid excrements are found all those substances taken in the food which have undergone no alteration in the digestive organs, all indigestible matters, such as woody fibre, the green colouring matter of leaves (chlorophyle), wax,

Physiology teaches us, that the process of nutrition in animals, that is, their increase of bulk, or the restoration of wasted parts, proceeds from the blood. The purpose of digestion and assimilation is to convert the food into blood. In the stomach and intestines, therefore, all those substances in the food capable of conversion into blood are separated from its other constituents; in other words, during the passage of the food through the intestinal canal there is a constant absorption of its nitrogen, since only azotised substances are capable of conversion into blood; and therefore the solid excrements are destitute of that element, except only a small portion, in the constitution of that secretion which is formed to facilitate their passage. With the solid excrements, the phosphates of lime and magnesia, which were contained in the food and not assimilated, are carried off, these salts being insoluble in water, and therefore not entering the urine.

We may obtain a clear insight into the chemical constitution of the solid excrements without further investigation, by comparing the faeces of a dog with his food. We give that animal flesh and bones – substances rich in azotised matter – and we obtain, as the last product of its digestion, a perfectly white excrement, solid while moist, but becoming in dry air a powder. This is the phosphate of lime of the bones, with scarcely one per cent. of foreign organic matter.

Thus we see that in the solid and fluid excrements of man and animals, all the nitrogen – in short, all the constituent ingredients of the consumed food, soluble and insoluble, are returned; and as food is primarily derived from the fields, we possess in those excrements all the ingredients which we have taken from it in the form of seeds, roots, or herbs.

One part of the crops employed for fattening sheep and cattle is consumed by man as animal food; another part is taken directly – as flour, potatoes, green vegetables, a third portion consists of vegetable refuse, and straw employed as litter. None of the materials of the soil need be lost. We can, it is obvious, get back all its constituent parts which have been withdrawn therefrom, as fruits, grain and animals, in the fluid and solid excrements of man, and the bones, blood and skins of the slaughtered animals. It depends upon ourselves to collect carefully all these scattered elements, and to restore the disturbed equilibrium of composition in the soil. We can calculate exactly how much and which of the component parts of the soil we export in a sheep or an ox, in a quarter of barley, wheat or potatoes, and we can discover, from the known composition of the excrements of man and animals, how much we have to supply to restore what is lost to our fields.

If, however, we could procure from other sources the substances which give to the exuviae of man and animals their value in agriculture, we should not need the latter. It is quite indifferent for our purpose whether we supply the ammonia (the source of nitrogen) in the form of urine, or in that of a salt derived from coal-tar; whether we derive the phosphate of lime from bones, apatite, or fossil excrements (the coprolithes).

The principal problem for agriculture is, how to replace those substances which have been taken from the soil, and which cannot be furnished by the atmosphere. If the manure supplies an imperfect compensation for this loss, the fertility of a field or of a country decreases; if, on the contrary, more are given to the fields, their fertility increases.

An importation of urine, or of solid excrements, from a foreign country, is equivalent to an importation of grain and cattle. In a certain time, the elements of those substances assume the form of grain, or of fodder, then become flesh and bones, enter into the human body, and return again day by day to the form they originally possessed.

The only real loss of elements we are unable to prevent is of the phosphates, and these, in accordance with the customs of all modern nations, are deposited in the grave. For the rest, every part of that enormous quantity of food which a man consumes during his lifetime (say in sixty or seventy years), which was derived from the fields, can be obtained and returned to them. We know with absolute certainty, that in the blood of a young or growing animal there remains a certain quantity of phosphate of lime and of the alkaline phosphates, to be stored up and to minister to the growth of the bones and general bulk of the body, and that, with the exception of this very small quantity, we receive back, in the solid and fluid excrements, all the salts and alkaline bases, all the phosphate of lime and magnesia, and consequently all the inorganic elements which the animal consumes in its food.

We can thus ascertain precisely the quantity, quality, and composition of animal excrements, without the trouble of analysing them. If we give a horse daily 4 1/2 pounds' weight of oats, and 15 pounds of hay, and knowing that oats give 4 per cent. and hay 9 per cent. of ashes, we can calculate that the daily excrements of the horse will contain 21 ounces of inorganic matter which was drawn from the fields. By analysis we can determine the exact relative amount of silica, of phosphates, and of alkalies, contained in the ashes of the oats and of the hay.

You will now understand that the constituents of the solid parts of animal excrements, and therefore their qualities as manure, must vary with the nature of the creature's food. If we feed a cow upon beetroot, or potatoes, without hay, straw or grain, there will be no silica in her solid excrements, but there will be phosphate of lime and magnesia. Her fluid excrements will contain carbonate of potash and soda, together with compounds of the same bases with inorganic acids. In one word, we have, in the fluid excrements, all the soluble parts of the ashes of the consumed food; and in the solid excrements, all those parts of the ashes which are insoluble in water.

If the food, after burning, leaves behind ashes containing soluble alkaline phosphates, as is the case with bread, seeds of all kinds, and flesh, we obtain from the animal by which they are consumed a urine holding in solution these phosphates. If, however, the ashes of food contain no alkaline phosphates, but abound in insoluble earthy phosphates, as hay, carrots, and potatoes, the urine will be free from alkaline phosphates, but the earthy phosphates will be found in the faeces. The urine of man, of carnivorous and graminivorous animals, contains alkaline phosphates; that of herbivorous animals is free from these salts.

The analysis of the excrements of man, of the piscivorous birds (as the guano), of the horse, and of cattle, furnishes us with the precise knowledge of the salts they contain, and demonstrates, that in those excrements, we return to the fields the ashes of the plants which have served as food, – the soluble and insoluble salts and earths indispensable to the development of cultivated plants, and which must be furnished to them by a fertile soil.

There can be no doubt that, in supplying these excrements to the soil, we return to it those constituents which the crops have removed from it, and we renew its capability of nourishing new crops: in one word, we restore the disturbed equilibrium; and consequently, knowing that the elements of the food derived from the soil enter into the urine and solid excrements of the animals it nourishes, we can with the greatest facility determine the exact value of the different kinds of manure. Thus the excrements of pigs which we have fed with peas and potatoes are principally suited for manuring crops of potatoes and peas. In feeding a cow upon hay and turnips, we obtain a manure containing the inorganic elements of grasses and turnips, and which is therefore preferable for manuring turnips. The excrement of pigeons contains the mineral elements of grain; that of rabbits, the

elements of herbs and kitchen vegetables. The fluid and solid excrements of man, however, contain the mineral elements of grain and seeds in the greatest quantity.

LETTER XV

My dear Sir,

You are now acquainted with my opinions respecting the effects of the application of mineral agents to our cultivated fields, and also the rationale of the influence of the various kinds of manures; you will, therefore, now readily understand what I have to say of the sources whence the carbon and nitrogen, indispensable to the growth of plants, are derived.

The growth of forests, and the produce of meadows, demonstrate that an inexhaustible quantity of carbon is furnished for vegetation by the carbonic acid of the atmosphere.

We obtain from an equal surface of forest, or meadowland, where the necessary mineral elements of the soil are present in a suitable state, and to which no carbonaceous matter whatever is furnished in manures, an amount of carbon, in the shape of wood and hay, quite equal, and oftimes more than is produced by our fields, in grain, roots, and straw, upon which abundance of manure has been heaped.

It is perfectly obvious that the atmosphere must furnish to our cultivated fields as much carbonic acid, as it does to an equal surface of forest or meadow, and that the carbon of this carbonic acid is assimilated, or may be assimilated by the plants growing there, provided the conditions essential to its assimilation, and becoming a constituent element of vegetables, exist in the soil of these fields.

In many tropical countries the produce of the land in grain or roots, during the whole year, depends upon one rain in the spring. If this rain is deficient in quantity, or altogether wanting, the expectation of an abundant harvest is diminished or destroyed.

Now it cannot be the water merely which produces this enlivening and fertilising effect observed, and which lasts for weeks and months. The plant receives, by means of this water, at the time of its first development, the alkalies, alkaline earths, and phosphates, necessary to its organization. If these elements, which are necessary previous to its assimilation of atmospheric nourishment, be absent, its growth is retarded. In fact, the development of a plant is in a direct ratio to the amount of the matters it takes up from the soil. If, therefore, a soil is deficient in these mineral constituents required by plants, they will not flourish even with an abundant supply of water.

The produce of carbon on a meadow, or an equal surface of forest land, is independent of a supply of carbonaceous manure, but it depends upon the presence of certain elements of the soil which in themselves contain no carbon, together with the existence of conditions under which their assimilation by plants can be effected. We increase the produce of our cultivated fields, in carbon, by a supply of lime, ashes, and marl, substances which cannot furnish carbon to the plants, and yet it is indisputable, – being founded upon abundant experience, – that in these substances we furnish to the fields elements which greatly increase the bulk of their produce, and consequently the amount of carbon.

If we admit these facts to be established, we can no longer doubt that a deficient produce of carbon, or in other words, the barrenness of a field does not depend upon carbonic acid, because we are able to increase the produce, to a certain degree, by a supply of substances which do not contain any carbon. The same source whence the meadow and the forest are furnished with carbon, is also open to our cultivated plants. The great object of agriculture, therefore, is to discover the means best adapted to enable these plants to assimilate the carbon of the atmosphere which exists in it as carbonic acid. In furnishing plants, therefore, with mineral elements, we give them the power to appropriate carbon from a source which is inexhaustible; whilst in the absence of these elements the most abundant supply of carbonic acid, or of decaying vegetable matter, would not increase the produce of a field.

With an adequate and equal supply of these essential mineral constituents in the soil, the amount of carbonic acid absorbed by a plant from the atmosphere in a given time is limited by the quantity which is brought into contact with its organs of absorption.

The withdrawal of carbonic acid from the atmosphere by the vegetable organism takes place chiefly through its leaves; this absorption requires the contact of the carbonic acid with their surface, or with the part of the plant by which it is absorbed.

The quantity of carbonic acid absorbed in a given time is in direct proportion to the surface of the leaves and the amount of carbonic acid contained in the air; that is, two plants of the same kind and the same extent of surface of absorption, in equal times and under equal conditions, absorb one and the same amount of carbon.

In an atmosphere containing a double proportion of carbonic acid, a plant absorbs, under the same condition, twice the quantity of carbon. Boussingault observed, that the leaves of the vine, inclosed in a vessel, withdrew all the carbonic acid from a current of air which was passed through it, however great its velocity. (Dumas Lecon, p.23.) If, therefore, we supply double the quantity of carbonic acid to one plant, the extent of the surface of which is only half that of another living in ordinary atmospheric air, the former will obtain and appropriate as much carbon as the latter. Hence results the effects of humus, and all decaying organic substances, upon vegetation. If we suppose all the conditions for the absorption of carbonic acid present, a young plant will increase in mass, in a limited time, only in proportion to its absorbing surface; but if we create in the soil a new source of carbonic acid, by decaying vegetable substances, and the roots absorb in the same time three times as much carbonic acid from the soil as the leaves derive from the atmosphere, the plant will increase in weight fourfold. This fourfold increase extends to the leaves, buds, stalks, and in the increased extent of the surface, the plant acquires an increased power of absorbing nourishment from the air, which continues in action far beyond the time when its derivation of carbonic acid through the roots ceases. Humus, as a source of carbonic acid in cultivated lands, is not only useful as a means of increasing the quantity of carbon – an effect which in most cases may be very indifferent for agricultural purposes – but the mass of the plant having increased rapidly in a short time, space is obtained for the assimilation of the elements of the soil necessary for the formation of new leaves and branches.

Water evaporates incessantly from the surface of the young plant; its quantity is in direct proportion to the temperature and the extent of the surface. The numerous radical fibrillae replace, like so many pumps, the evaporated water; and so long as the soil is moist, or penetrated with water, the indispensable elements of the soil, dissolved in the water, are supplied to the plant. The water absorbed by the plant evaporating in an aeriform state leaves the saline and other mineral constituents within it. The relative proportion of these elements taken up by a plant, is greater, the more extensive the surface and more abundant the supply of water; where these are limited, the plant soon reaches its full growth, while if their supply is continued, a greater amount of elements necessary to enable it to appropriate atmospheric nourishment being obtained, its development proceeds much further. The quantity, or mass of seed produced, will correspond to the quantity of mineral constituents present in the plant. That plant, therefore, containing the most alkaline phosphates and earthy salts will produce more or a greater weight of seeds than another which, in an equal time has absorbed less of them. We consequently observe, in a hot summer, when a further supply of mineral ingredients from the soil ceases through want of water, that the height and strength of plants, as well as the development of their seeds, are in direct proportion to its absorption of the elementary parts of the soil in the preceding epochs of its growth.

The fertility of the year depends in general upon the temperature, and the moisture or dryness of the spring, if all the conditions necessary to the assimilation of the atmospheric nourishment be secured to our cultivated plants. The action of humus, then, as we have explained it above, is chiefly of value in gaining time. In agriculture, this must ever be taken into account and in this respect humus is of importance in favouring the growth of vegetables, cabbages,

But the cerealia, and plants grown for their roots, meet on our fields, in the remains of the preceding crop, with a quantity of decaying vegetable substances corresponding to their contents of mineral nutriment from the soil, and consequently with a quantity of carbonic acid adequate to their accelerated development in the spring. A further supply of carbonic acid, therefore, would be quite useless, without a corresponding increase of mineral ingredients.

From a morgen of good meadow land, 2,500 pounds weight of hay, according to the best agriculturists, are obtained on an average. This amount is furnished without any supply of organic substances, without manure containing carbon or nitrogen. By irrigation, and the application of ashes or gypsum, double that amount may be grown. But assuming 2,500 pounds weight of hay to be the maximum, we may calculate the amount of carbon and nitrogen derived from the atmosphere by the plants of meadows.

According to elementary analysis, hay, dried at a temperature of 100 deg Reaumur, contains 45.8 per cent. of carbon, and 1 1/2 per cent. of nitrogen. 14 per cent. of water retained by the hay, dried at common temperatures, is driven off at 100 deg. 2,500 pounds weight of hay, therefore, corresponds to 2,150 pounds, dried at 100 deg. This shows us, that 984 pounds of carbon, and 32.2 pounds weight of nitrogen, have been obtained in the produce of one morgen of meadow land. Supposing that this nitrogen has been absorbed by the plants in the form of ammonia, the atmosphere contains 39.1 pounds weight of ammonia to every 3640 pounds weight of carbonic acid (=984 carbon, or 27 per cent.), or in other words, to every 1,000 pounds weight of carbonic acid, 10.7 pounds of ammonia, that is to about 1/100,000, the weight of the air, or 1/60,000 of its volume.

For every 100 parts of carbonic acid absorbed by the surface of the leaves, the plant receives from the atmosphere somewhat more than one part of ammonia.

With every 1,000 pounds of carbon, we obtain—

From a meadow . 32 7/10 pounds of nitrogen.

From cultivated fields,

“In Wheat . 21 1/2”

“Oats . 22.3”

“Rye . 15.2”

“Potatoes . 34.1”

“Beetroot . 39.1”

“Clover . 44”

"Peas . 62"

Boussingault obtained from his farm at Bechelbronn, in Alsace, in five years, in the shape of potatoes, wheat, clover, turnips, and oats, 8,383 of carbon, and 250.7 nitrogen. In the following five years, as beetroot, wheat, clover, turnips, oats, and rye, 8,192 of carbon, and 284.2 of nitrogen. In a further course of six years, potatoes, wheat, clover, turnips, peas, and rye, 10,949 of carbon, 356.6 of nitrogen. In 16 years, 27,424 carbon, 858 1/2 nitrogen, which gives for every 1,000 carbon, 31.3 nitrogen.

From these interesting and unquestionable facts, we may deduce some conclusions of the highest importance in their application to agriculture.

1. We observe that the relative proportions of carbon and nitrogen, stand in a fixed relation to the surface of the leaves. Those plants, in which all the nitrogen may be said

to be concentrated in the seeds, as the cerealia, contain on the whole less nitrogen than the leguminous plants, peas, and clover.

2. The produce of nitrogen on a meadow which receives no nitrogenised manure, is greater than that of a field of wheat which has been manured.

3. The produce of nitrogen in clover and peas, which agriculturists will acknowledge require no nitrogenised manure, is far greater than that of a potato or turnip field, which is abundantly supplied with such manures.

Lastly. And this is the most curious deduction to be derived from the above facts, – if we plant potatoes, wheat, turnips, peas, and clover, (plants containing potash, lime, and silex,) upon the same land, three times manured, we gain in 16 years, for a given quantity of carbon, the same proportion of nitrogen which we receive from a meadow which has received no nitrogenised manure.

On a morgen of meadowland, we obtain in plants, containing silex, lime, and potash, 984 carbon, 32.2 nitrogen. On a morgen of cultivated land, in an average of 16 years, in plants containing the same mineral elements, silex, lime, and potash, 857 carbon, 26.8 nitrogen.

If we add the carbon and nitrogen of the leaves of the beetroot, and the stalk and leaves of the potatoes, which have not been taken into account, it still remains evident that the cultivated fields, notwithstanding the supply of carbonaceous and nitrogenised manures, produced no more carbon and nitrogen than an equal surface of meadowland supplied only with mineral elements.

What then is the rationale of the effect of manure, – of the solid and fluid excrements of animals?

This question can now be satisfactorily answered: that effect is the restoration of the elementary constituents of the soil which have been gradually drawn from it in the shape of grain and cattle. If the land I am speaking of had not been manured during those 16 years, not more than one-half, or perhaps than one-third part of the carbon and nitrogen would have been produced. We owe it to the animal excrements, that it equalled in production the meadowland, and this, because they restored the mineral ingredients of the soil removed by the crops. All that the supply of manure accomplished, was to prevent the land from becoming poorer in these, than the meadow which produces 2,500 pounds of hay. We withdraw from the meadow in this hay as large an amount of mineral substances as we do in one harvest of grain, and we know that the fertility of the meadow is just as dependent upon the restoration of these ingredients to its soil, as the cultivated land is upon manures. Two meadows of equal surface, containing unequal quantities of inorganic elements of nourishment, – other conditions being equal, – are very unequally fertile; that which possesses most, furnishes most hay. If we do not restore to a meadow the withdrawn elements, its fertility decreases. But its fertility remains unimpaired, with a due supply of animal excrements, fluid and solid, and it not only remains the same, but may be increased by a supply of mineral substances alone, such as remain after the combustion of ligneous plants and other vegetables; namely, ashes. Ashes represent the whole nourishment which vegetables receive from the soil. By furnishing them in sufficient quantities to our meadows, we give to the plants growing on them the power of condensing and absorbing carbon and nitrogen by their surface. May not the effect of the solid and fluid excrements, which are the ashes of plants and grains, which have undergone combustion in the bodies of animals and of man, be dependent upon the same cause? Should not the fertility, resulting from their application, be altogether independent of the ammonia they contain? Would not their effect be precisely the same in promoting the fertility of cultivated plants, if we had evaporated the urine, and dried and burned the solid excrements? Surely the cerealia and leguminous plants which we cultivate must derive their carbon and nitrogen from the same source whence the graminea and leguminous plants of the meadows obtain them! No doubt can be entertained of their capability to do so.

In Virginia, upon the lowest calculation, 22 pounds weight of nitrogen were taken on the average, yearly, from every morgen of the wheat-fields. This would amount, in 100 years, to 2,200 pounds weight. If this were derived from the soil, every morgen of it must have contained the equivalent of 110,000 pounds weight of animal excrements (assuming the latter, when dried, at the temperature of boiling water, to contain 2 per cent.).

In Hungary, as I remarked in a former Letter, tobacco and wheat have been grown upon the same field for centuries, without any supply of nitrogenised manure. Is it possible that the nitrogen essential to, and entering into, the composition of these crops, could have been drawn from the soil?

Every year renews the foliage and fruits of our forests of beech, oak, and chesnuts; the leaves, the acorns, the chesnuts, are rich in nitrogen; so are cocoa-nuts, breadfruit, and other tropical productions. This nitrogen is not supplied by man, can it indeed be derived from any other source than the atmosphere?

In whatever form the nitrogen supplied to plants may be contained in the atmosphere, in whatever state it may be when absorbed, from the atmosphere it must have been derived. Did not the fields of Virginia receive their nitrogen from the same source as wild plants?

Is the supply of nitrogen in the excrements of animals quite a matter of indifference, or do we receive back from our fields a quantity of the elements of blood corresponding to this supply?

The researches of Boussingault have solved this problem in the most satisfactory manner. If, in his grand experiments, the manure which he gave to his fields was in the same state, i.e. dried at 110 deg in a vacuum, as it was when analysed, these fields received, in 16 years, 1,300 pounds of nitrogen. But we know that by drying all the nitrogen escapes which is contained in solid animal excrements, as volatile carbonate of ammonia. In this calculation the nitrogen of the urine, which by decomposition is converted into carbonate of ammonia, has not been included. If we suppose it amounted to half as much as that in the dried excrements, this would make the quantity of nitrogen supplied to the fields 1,950 pounds.

In 16 years, however, as we have seen, only 1,517 pounds of nitrogen, was contained in their produce of grain, straw, roots, et cetera

– that is, far less than was supplied in the manure; and in the same period the same extent of surface of good meadowland (one hectare = a Hessian morgen), which received no nitrogen in manure, 2,062 pounds of nitrogen.

It is well known that in Egypt, from the deficiency of wood, the excrement of animals is dried, and forms the principal fuel, and that the nitrogen from the soot of this excrement was, for many centuries, imported into Europe in the form of sal ammoniac, until a method of manufacturing this substance was discovered at the end of the last century by Gravenhorst of Brunswick. The fields in the delta of the Nile are supplied with no other animal manures than the ashes of the burnt excrements, and yet they have been proverbially fertile from a period earlier than the first dawn of history, and that fertility continues to the present day as admirable as it was in the earliest times. These fields receive, every year, from the inundation of the Nile, a new soil, in its mud deposited over their surface, rich in those mineral elements which have been withdrawn by the crops of the previous harvest. The mud of the Nile contains as little nitrogen as the mud derived from the Alps of Switzerland, which fertilises our fields after the inundations of the Rhine. If this fertilising mud owed this property to nitrogenised matters; what enormous beds of animal and vegetable exuviae and remains ought to exist in the mountains of Africa, in heights extending beyond the limits of perpetual snow, where no bird, no animal finds food, from the absence of all vegetation!

Abundant evidence in support of the important truth we are discussing, may be derived from other well known facts. Thus, the trade of Holland in cheese may be adduced in proof and illustration thereof. We know that cheese is derived from the plants which serve as food for cows. The meadowlands of Holland derive the nitrogen of cheese from the same source as with us; i.e. the atmosphere. The milch cows of Holland remain day and night on the grazing-grounds, and therefore, in their fluid and solid excrements return directly to the soil all the salts and earthy elements of their food: a very insignificant quantity only is exported in the cheese. The fertility of these meadows can, therefore, be as little impaired as our own fields, to which we restore all the elements of the soil, as manure, which have been withdrawn in the crops. The only difference is, in Holland they remain on the field, whilst we collect them at home and carry them, from time to time, to the fields.

The nitrogen of the fluid and solid excrements of cows, is derived from the meadow-plants, which receive it from the atmosphere; the nitrogen of the cheese also must be drawn from the same source. The meadows of Holland have, in the lapse of centuries, produced millions of hundredweights of cheese. Thousands of hundredweights are annually exported, and yet the productiveness of the meadows is in no way diminished, although they never receive more nitrogen than they originally contained.

Nothing then can be more certain than the fact, that an exportation of nitrogenised products does not exhaust the fertility of a country; inasmuch as it is not the soil, but the atmosphere, which furnishes its vegetation with nitrogen. It follows, consequently, that we cannot increase the fertility of our fields by a supply of nitrogenised manure, or by salts of ammonia, but rather that their produce increases or diminishes, in a direct ratio, with the supply of mineral elements capable of assimilation. The formation of the constituent elements of blood, that is, of the nitrogenised principles in our cultivated plants, depends upon the presence of inorganic matters in the soil, without which no nitrogen can be assimilated even when there is a most abundant supply. The ammonia contained in animal excrements exercises a favourable effect, inasmuch as it is accompanied by the other substances necessary to accomplish its transition into the elements of the blood. If we supply ammonia associated with all the conditions necessary to its assimilation, it ministers to the nourishment of the plants; but if this artificial supply is not given they can derive all the needed nitrogen from the atmosphere – a source, every loss from which is restored by the decomposition of the bodies of dead animals and the decay of plants. Ammonia certainly favours, and accelerates, the growth of plants in all soils, wherein all the conditions of its assimilation are united; but it is altogether without effect, as respects the production of the elements of blood where any of these conditions are wanting. We can suppose that asparagin, the active constituent of asparagus, the mucilaginous root of the marshmallow, the nitrogenised and sulphurous ingredients of mustard-seed, and of all cruciferous plants, may originate without the aid of the mineral elements of the soil. But if the principles of those vegetables, which serve as food, could be generated without the co-operation of the mineral elements of blood, without potash, soda, phosphate of soda, phosphate of lime, they would be useless to us and to herbivorous animals as food; they would not fulfil the purpose for which the wisdom of the Creator has destined them. In the absence of alkalies and the phosphates, no blood, no milk, no muscular fibre can be formed. Without phosphate of lime our horses, sheep and cattle, would be without bones.

In the urine and in the solid excrements of animals we carry ammonia, and, consequently, nitrogen, to our cultivated plants, and this nitrogen is accompanied by all the mineral elements of food exactly in the same proportions, in which both are contained in the plants which served as food to the animals, or what is the same, in those proportions in which both can serve as nourishment to a new generation of plants, to which both are essential.

The effect of an artificial supply of ammonia, as a source of nitrogen, is, therefore, precisely analogous to that of humus as a source of carbonic acid – it is limited to a gain of time; that is, it accelerates the development of plants. This is of great importance, and should always be taken into account in gardening, especially in the treatment of the kitchen-garden; and as much as possible, in agriculture on a large scale, where the time occupied in the growth of the plants cultivated is of importance.

When we have exactly ascertained the quantity of ashes left after the combustion of cultivated plants which have grown upon all varieties of soil, and have obtained correct analyses of these ashes, we shall learn with certainty which of the constituent elements of the plants are constant and which are changeable, and we shall arrive at an exact knowledge of the sum of all the ingredients we withdraw from the soil in the different crops.

With this knowledge the farmer will be able to keep an exact record, of the produce of his fields in harvest, like the account-book of a well regulated manufactory; and then by a simple calculation he can determine precisely the substances he must supply to each field, and the quantity of these, in order to restore their fertility. He will be able to express, in pounds weight, how much of this or that element he must give in order to augment its fertility for any given kind of plants.

These researches and experiments are the great desideratum of the present time. TO THE UNITED EFFORTS OF THE CHEMISTS OF ALL COUNTRIES WE MAY CONFIDENTLY LOOK FOR A SOLUTION OF THESE GREAT QUESTIONS, and by the aid of ENLIGHTENED AGRICULTURISTS we shall arrive at a RATIONAL system of GARDENING, HORTICULTURE, and AGRICULTURE, applicable to every country

and all kinds of soil, and which will be based upon the immutable foundation of OBSERVED FACTS and PHILOSOPHICAL INDUCTION.

LETTER XVI

My dear Sir,

My recent researches into the constituent ingredients of our cultivated fields have led me to the conclusion that, of all the elements furnished to plants by the soil and ministering to their nourishment, the phosphate of lime – or, rather, the phosphates generally – must be regarded as the most important.

In order to furnish you with a clear idea of the importance of the phosphates, it may be sufficient to remind you of the fact, that the blood of man and animals, besides common salt, always contains alkaline and earthy phosphates. If we burn blood and examine the ashes which remain, we find certain parts of them soluble in water, and others insoluble. The soluble parts are, common salt and alkaline phosphates; the insoluble consist of phosphate of lime, phosphate of magnesia, and oxide of iron.

These mineral ingredients of the blood – without the presence of which in the food the formation of blood is impossible – both man and animals derive either immediately, or mediately through other animals, from vegetable substances used as food; they had been constituents of vegetables, they had been parts of the soil upon which the vegetable substances were developed.

If we compare the amount of the phosphates in different vegetable substances with each other, we discover a great variety, whilst there is scarcely any ashes of plants altogether devoid of them, and those parts of plants which experience has taught us are the most nutritious, contain the largest proportion. To these belong all seeds and grain, especially the varieties of bread-corn, peas, beans, and lentils.

It is a most curious fact that if we incinerate grain or its flour, peas, beans, and lentils, we obtain ashes, which are distinguished from the ashes of all other parts of vegetables by the absence of alkaline carbonates. The ashes of these seeds when recently prepared, do not effervesce with acids; their soluble ingredients consist solely of alkaline phosphates, the insoluble parts of phosphate of lime, phosphate of magnesia, and oxide of iron: consequently, of the very same salts which are contained in blood, and which are absolutely indispensable to its formation. We are thus brought to the further indisputable conclusion that no seed suitable to become food for man and animals can be formed in any plant without the presence and co-operation of the phosphates. A field in which phosphate of lime, or the alkaline phosphates, form no part of the soil, is totally incapable of producing grain, peas, or beans.

An enormous quantity of these substances indispensable to the nourishment of plants, is annually withdrawn from the soil and carried into great towns, in the shape of flour, cattle, et cetera. It is certain that this incessant removal of the phosphates must tend to exhaust the land and diminish its capability of producing grain. The fields of Great Britain are in a state of progressive exhaustion from this cause, as is proved by the rapid extension of the cultivation of turnips and mangel wurzel – plants which contain the least amount of the phosphates, and therefore require the smallest quantity for their development. These roots contain 80 to 92 per cent. of water. Their great bulk makes the amount of produce fallacious, as respects their adaptation to the food of animals, inasmuch as their contents of the ingredients of the blood, i.e. of substances which can be transformed into flesh, stands in a direct ratio to their amount of phosphates,

without which neither blood nor flesh can be formed.

Our fields will become more and more deficient in these essential ingredients of food, in all localities where custom and habits do not admit the collection of the fluid and solid excrements of man, and their application to the purposes of agriculture. In a former letter I showed you how great a waste of phosphates is unavoidable in England, and referred to the well-known fact that the importation of bones restored in a most admirable manner the fertility of the fields exhausted from this cause. In the year 1827 the importation of bones for manure amounted to 40,000 tons, and Huskisson estimated their value to be from L 100,000 to L 200,000 sterling. The importation is still greater at present, but it is far from being sufficient to supply the waste.

Another proof of the efficacy of the phosphates in restoring fertility to exhausted land is afforded by the use of the guano – a manure which, although of recent introduction into England, has found such general and extensive application.

We believe that the importation of one hundredweight of guano is equivalent to the importation of eight hundredweight of wheat – the hundredweight of guano assumes in a time which can be accurately estimated the form of a quantity of food corresponding to eight hundredweight of wheat. The same estimate is applicable in the valuation of bones.

If it were possible to restore to the soil of England and Scotland the phosphates which during the last fifty years have been carried to the sea by the Thames and the Clyde, it would be equivalent to manuring with millions of hundredweights of bones, and the produce of the land would increase one-third, or perhaps double itself, in five to ten years.

We cannot doubt that the same result would follow if the price of the guano admitted the application of a quantity to the surface of the fields, containing as much of the phosphates as have been withdrawn from them in the same period.

If a rich and cheap source of phosphate of lime and the alkaline phosphates were open to England, there can be no question that the importation of foreign corn might be altogether dispensed with after a short time. For these materials England is at present dependent upon foreign countries, and the high price of guano and of bones prevents their general application, and in sufficient quantity. Every year the trade in these substances must decrease, or their price will rise as the demand for them increases.

According to these premises, it cannot be disputed, that the annual expense of Great Britain for the importation of bones and guano is equivalent to a duty on corn: with this difference only, that the amount is paid to foreigners in money.

To restore the disturbed equilibrium of constitution of the soil, – to fertilise her fields, – England requires an enormous supply of animal excrements, and it must, therefore, excite considerable interest to learn, that she possesses beneath her soil beds of fossil guano, strata of animal excrements, in a state which will probably allow of their being employed as a manure at a very small expense. The coprolithes discovered by Dr. Buckland, (a discovery of the highest interest to Geology,) are these excrements; and it seems extremely probable that in these strata England possesses the means of supplying the place of recent bones, and therefore the principal conditions of improving agriculture – of restoring and exalting the fertility of her fields.

In the autumn of 1842, Dr. Buckland pointed out to me a bed of coprolithes in the neighbourhood of Clifton, from half to one foot thick, inclosed in a limestone formation, extending as a brown stripe in the rocks, for miles along the banks of the Severn. The limestone marl of Lyme Regis consists, for the most part, of one-fourth part of fossil excrements and bones. The same are abundant in the lias of Bath, Eastern and Broadway Hill, near Evesham. Dr. Buckland mentions beds, several miles in extent, the substance of which consists, in many places, of a fourth part of coprolithes.

Pieces of the limestone rock in Clifton, near Bristol, which is rich in coprolithes and organic remains, fragments of bones, teeth, were subjected to analysis, and were found to contain above 18 per cent. of phosphate of lime. If this limestone is burned and brought in that state to the fields, it must be a perfect substitute for bones, the efficacy of which as a manure does not depend, as has been generally, but erroneously supposed, upon the nitrogenised matter which they contain, but on their phosphate of lime.

The osseous breccia found in many parts of England deserves especial attention, as it is highly probable that in a short time it will become an important article of commerce.

What a curious and interesting subject for contemplation! In the remains of an extinct animal world, England is to find the means of increasing her wealth in agricultural produce, as she has already found the great support of her manufacturing industry in fossil fuel, – the preserved matter of primeval forests, – the remains of a vegetable world. May this expectation be realised! and may her excellent population be thus redeemed from poverty and misery!

Your Customized Book Source

ReadHowYouWant has developed innovative, patent-pending technology that enables us to customize books to suit individual readers.

We continually create new editions that make reading easier, quicker and more enjoyable for readers. Our EasyRead printed books have been optimized to improve word recognition and ease eye tracking by adjusting word and line spacing and minimizing hyphenation.

In addition to several larger-font formats, we offer formats for people with dyslexia, macular degeneration, and eye-tracking problems. We also produce Braille and DAISY formats and can custom-build a text that is perfect for you.

These editions are available at ReadHowYouWant.com/bookshop or at Amazon.com. For more information please visit ReadHowYouWant.com/format

To find your custom format, see the next page.

The EasyRead Edition is optimized for reading enjoyment for readers with normal vision.

The EasyRead Comfort Edition is optimized for readers who find that reading small print requires effort and eye focus but do not need large type.

The EasyRead Large Edition is optimized for readers who need a standard 16pt large type.

The EasyRead Large Bold Edition is optimized for readers with reduced vision who need intense bold 16pt large type.

The EasyRead Super Large 18 Edition is optimized for readers with significantly reduced vision who need intense bold 18pt large type.

The EasyRead Super Large 20 Edition is optimized for readers with significantly reduced vision who need intense bold 20pt large type.

The EasyRead Super Large 24 Edition is optimized for readers with severely reduced vision who need intense bold 24pt large type.

The EasyRead **Word** Pattern **Edition** is **optimized** for **readers** who **need** help **differentiating** between **words** or **to** track **their** eyes **across** a **line.** This

pattern adds **extra** visual **information** about **word** order, **and** makes **it** harder **for** readers **to** swap **the** order **of** adjacent **words.**

The EasyRead Character Pattern Edition. This edition makes it difficult to forget the direction of reading. It makes the ending of the word stand out, and this help people who forget to read the last syllable of the word. There is extra visual information about the order of letters within the word, and the pattern can also help with word differentiation and eye tracking.

The EasyRead Custom Character Discrimination Pattern distinguishes between letters by making them stand out from similarly shaped letters: e and **o**, mirrored letters like b and **d**, or stemmed and unstemmed letters like h and **n.** This format is offered as a Custom Format as the character discrimination issues faced by a reader are usually specific to that individual.

The EasyRead Custom format allows readers to select the format that's customized to their own needs. Readers can select typeface, font size, character, word, and line spacing, as well as line length. The other special formats can be used as options. Type color can also be selected.